THE MAN WHO PLANTED HIS OWN JUNGLE

Nelson "Omar Tello" Benalcázar

My hands will write what my eyes have seen, and my heart will declare what nature has taught me

Gran Camino Press S.A.S.

Memories Collection

THE MAN WHO PLANTED HIS OWN JUNGLE

Nelson "Omar Tello" Benalcázar, Author

Jon Jaramillo, Editor and Translator

Published by:
Gran Camino Press
2544 Washington Street
Eugene, Oregon 97405 USA

Typesetting: Jon Jaramillo

Cover Design: Jon Jaramillo

A CIP record for this book is available from the Library of Congress Cataloging-in-Publication Data

ISBN-13: 979-8-9920538-9-0 (Hardback)

ISBN-13: 979-8-9920538-8-3 (Paperback)

Printed in USA

EDITOR AND TRANSLATOR'S NOTE

I met Mr. Omar Tello in July 2025, during a Zoom conversation that moved me deeply. Although I had already visited his Las Orquídeas Botanical Garden and walked with amazement through the regenerated forest alongside his daughter Ángela, it was in that virtual meeting where I understood with complete clarity the dimension of his story: not just that of a naturalist or an environmental defender, but that of a man who has fought for five decades to protect what others insisted on destroying.

From the age of 3, Omar witnessed the arrival of outsiders to the Ecuadorian Amazon, attracted by the unparalleled wealth of its biodiversity. What was a paradise to him was merely plunder to others: thousands of species hunted, logged, or extracted as "trophies" without conscience or return. That childhood as a silent witness marked him forever and sowed in him the seed of an active resistance that would take shape over time.

As the years passed, he bought degraded lands and began to sow life where others only saw cattle pasture. He did so without state support, facing bureaucratic hurdles that for decades tried to suffocate his dream. And yet, he persevered. What was once devastated terrain is today a vibrant ecosystem: a refuge for wildlife, orchids, bromeliads, native trees, and ancestral knowledge that still resists oblivion.

But his struggle does not end there. Omar Tello has written a brave manuscript in which he not only tells his personal story but also develops a viable and replicable methodological proposal to restore what can still be saved. His text is, at the same time, a testimony, a well-founded critique against extractivism, and a roadmap to recover the social and cultural fabric that depends on the rainforest.

A COMPREHENSIVE TRANSLATION EFFORT

From GRANCAMINOPRESS S.A.S., it is an honor to bring this work to an English-speaking audience, amplifying a text that serves as a fundamental contribution to the ecological and cultural heritage of Ecuador. Our task was not limited to simply converting words from one language to another; we collaborated closely with Mr.

Tello to ensure that his voice, authentic and testimonial, remained intact in English, exactly as he conceived it from the rainforest.

Spanish *selva* has a semantic range that English divides among *rainforest*, *forest*, and *jungle*. In descriptive and ecological passages, *rainforest* or *forest* may be the most precise English equivalent; *jungle* is retained where its emotional, symbolic, or literary force best conveys Omar's conception of the *selva*, including in the title.

The translation process focused on three crucial aspects:

Clarity and Coherence: We carefully rendered the English manuscript to ensure that the reader can transition smoothly from the Remembrances (testimony) to the Methodology (technical proposal). It was vital to convey the complex connections between the ecological crisis and cultural dispossession without losing the rhythm of the original Spanish.

Contextualizing the Testimony: We adapted the critical apparatus that accompanies the narrative, ensuring that the academic references and nuances contextualizing Omar Tello's personal experience resonate effectively within the broader global debates on environmental justice, rights of nature, and Amazonian politics. To this end, readers will encounter the term "colonizer," which may suggest Spanish colonial conquest. The term in Tello's context refers to Ecuadorian migrants encouraged by twentieth-century state colonization policies to settle Amazonian territory, many of whom were poor Andean farming families. We also translate the term "campesino" as "peasant," "farmer," or "peasant farmer" according to the context.

The Soul of the Author: We meticulously preserved the direct language, the cadence, and the intentional linguistic choices that make this book a unique tale of ecological memory. Our goal was to ensure the English version captures the true spirit and celebration of the Amazon's resilience.

In these pages, there is wisdom, there is history, and there is an urgent call: if we listen closely to the voice of this man who planted his own rainforest, we still have time to respond.

A NOTE ABOUT THE ENGLISH EDITION

This English edition of Tello's text does not attempt to recast the author as a scientist he never claimed to be. Rather, it presents more than forty years of field observation, a methodology grounded in that experience, and ecological processes that, in many instances, are independently supported by contemporary science. Historical events, institutional encounters, environmental changes, and interpretations of their causes are presented as Tello remembers and experienced them, preserving the testimonial character of the original work.

–Jon Dell Jaramillo, Editor and Translator

DEDICATION

This book is dedicated principally to the youth and to the children, for in them the hope of future defenders of life germinates.

I also dedicate it to my family: to my wife Lupe and my four children, Darwin, Jhoana, Ángela, and Carla, who endured the great sacrifices that the execution of this research demanded. A special dedication to my children Darwin and Carlita, who, with their special abilities, teach me every day about true strength and inspire me to continue.

I extend this dedication to society, with the hope of reaching the heart of every reader, so that this experience might be multiplied in many lives.

My memories go back to 1960, when I was barely four years old. Since then, my story has been interwoven with the research and the effort to regenerate ecosystems, a task I began in 1980 and that today materializes in 6.75 hectares of land that were once pasture but are now a living center for research.

This space, transformed into a jungle, demonstrates that it is possible to recover what was lost. Within it breathe the biodiversity indicators that sustain life, an indispensable heritage for humanity. Although the destruction has been ferocious, I hold to the certainty that if in our lifetime we rebuild what has been damaged and protect what still exists, hope will remain firm.

If we fulfill this mission, we will receive the reward of seeing life reborn. If not, it will be nature herself who demands it from us.

The paradise I observed as a child is no longer there. It is vital that we do not let our hopes and attitudes waver in the work of seeing the forest recover and be protected. Someone, possibly, planted it, and we had no right to use it so massively, because it is a substantial heritage of creation, which nourishes every being in the world with life—with oxygen, water, food, medicine, and innumerable benefits. If we conserve nature, even if we die, it will be with the satisfaction that a dignified life remains possible for future generations. Even if the power of destruction advances, we triumph all the same, because while we lived, we improved our surroundings and saved many lives. In that sense, nothing has been lost.

ACKNOWLEDGMENTS

To God, for having given me life and the gift of loving nature, for guiding me to connect with her, to overcome difficult moments, and for showing me the path to always find the most just solutions.

To my family, a constant source of inspiration and tenderness, like the very beings of the forest. My children, whom I watched grow like seeds, are today like sturdy trees. Together with this great nature, they represent the strongest bulwark, one that embraces even the smallest plant with its roots.

To my brother Efraín, who selflessly took part in numerous tasks of ecosystem reconstruction. To my friends, who directly or indirectly have supported me in different stages of this journey.

A very special thanks to Hannelore Seebass and Dr. Ernesto Seebass, whose extraordinary backing made possible the advancement of the environmental project, the well-being of my family, and the creation of a park accessible to children with disabilities. Without their support, none of this would have been possible.

TABLE OF CONTENTS

TABLE OF FIGURES

PROLOGUE

ACADEMIC AND ECOLOGICAL SIGNIFICANCE OF OMAR TELLO'S CONTRIBUTION

Omar Tello's book, *The Man Who Planted His Own Jungle*, stands out not only for its originality but, above all, for its significant contribution to the restoration of degraded areas in the Ecuadorian Amazon. It is the result of the author's systematic effort over several decades to restore 7 hectares of completely degraded land on the outskirts of El Puyo, transforming it into its current state, which bears characteristics similar to a primary forest. This unique and successful effort possesses several characteristics that make it an exceptional experience.

First, its author, who comes from a peasant farmer background and lacks specialized university training, has achieved these almost unique results in the Amazon region based on his profound empirical knowledge of the soils, the species he progressively introduced, and nature's responses. The experience thus becomes a living example of the necessary complementarity between the contributions of conventional science and the Indigenous and popular knowledge that has nourished Omar Tello. Using specialized terminology, this is a successful experience in transdisciplinarity.

A relevant aspect of Tello's work is the rigorous systematization of his experience. The main steps of this restoration process, which began in 1980, are clearly documented and supported by photographs, rendering the entire experience replicable.

Finally, the most important feature of Tello's contribution is its capacity for replication on a larger scale. As will be explained in the following paragraphs, the need to restore deforested and degraded areas in the Ecuadorian Amazon—while simultaneously recovering its biodiversity, its carbon sequestration capacity, and its contribution to rainfall—is urgent. For this effort to be effective, it is necessary to expand the scale of its application as much as possible.

The areas of highest biodiversity in the international Amazon basin are located in Ecuador, including the Yasuní National Park, the Cuyabeno Wildlife Reserve, and the basins of the Napo, Curaray, Pastaza, and Bobonaza rivers (Bass et al. 2010). This heritage, however, is seriously threatened. By 2018, the Ecuadorian Amazon had lost 13% of its primary forest, and a similar percentage was degraded (Science Panel for the Amazon 2021). Among the countries that make up the Amazon basin, only Brazil has a higher accumulated deforestation.

The combined effects of climate change, deforestation, and degradation generate a sustained loss of rainfall. This has manifested in prolonged droughts, such as the one that affected the country and the entire Amazon basin in 2024, resulting in widespread power outages throughout the country for several consecutive hours a day.

If deforestation continues at its current high levels without effective control, or if it expands, the current tree cover of the jungle may begin to disappear due to a lack of moisture, and the jungle could progressively be replaced by a savanna with scattered trees and minimal biodiversity (Wunderling et al. 2026). This process could be triggered in several areas of the Amazon basin, including Ecuador, and would lead to a severe expansion of droughts in the Andean highlands (Sierra)—whose water supply depends largely on the Amazon—and in cities like Quito. The economic and social consequences of the lack of rainfall will be severe, particularly for small-scale producers in the Sierra and the Amazon.

The two most effective strategies to prevent this future climate deterioration are reducing deforestation and expanding the restoration of degraded areas in the Amazon. Unfortunately, deforestation continues almost unchecked in Ecuador; between 2001 and 2025, the country lost 1,100,000 hectares, while recovery due to reforestation or spontaneous regeneration has barely reached 220,000 hectares (Global Forest Watch 2026).

Omar Tello's book provides us with an invaluable tool to expand the scale of regeneration of deforested areas in the Ecuadorian jungle—thus protecting the climate and biodiversity—based on a proven and extensively documented experience.

This regeneration of primary forests is necessary because the vast majority of Ecuadorian Amazon soils have a low nutrient content, as these are washed away by rain or rapidly degraded by high temperatures. A substantial proportion of the ecosystem's actively cycling nutrients is held in vegetation and rapidly recycled organic matter; once the vegetation is removed, the remaining nutrients in the soil can be depleted after just a few years of cultivation in deforested areas. As a result, agricultural productivity declines rapidly until it almost disappears, and the pasture cycle follows a similar trajectory.

Once the deforested plots lose their agricultural or livestock-raising capacity, they are abandoned, and their owners move on, deforesting new areas. This process occurs approximately 10 to 15 years after the initial deforestation, depending on specific conditions.

In this way, many degraded areas appear where the recovery method proposed by Omar Tello becomes highly effective. If an appropriate recovery strategy is not applied, a slow, spontaneous transition to secondary forests begins, but the recovery of biodiversity is slow and partial, and the time required for complete regeneration is much greater.

In summary, Omar Tello's book stands out not only for its originality but also for proposing proven solutions whose large-scale application can forge pathways to overcome the critical environmental problems facing the Ecuadorian Amazon.

—Dr. Carlos Larrea

June 2026

AUTHOR'S INTRODUCTION

I wrote this book in simple and understandable language because I want the reader not only to read but to know that this story lives. I want you to touch, feel, hear, and breathe within it. I refer to the nature that has accompanied me with patience and passion throughout my entire life. It led me to devise a methodology for reconstructing ecosystems—a method we can apply in various places to prevent the extinction of flora and fauna. This method always champions the concept of freedom. It neutralizes the successive destruction of nature in small or large environments and blocks the path toward eternal bondage.

Climate change—and the consequences people worldwide have discussed for decades—has intensified our concern. Yet, we have failed to take truly effective global measures to remedy it. However, committed groups and individuals exist. Though small against the power of the destroyers, they contribute concrete actions. I have dedicated my life to this research, working through errors and successes until I consolidated an effective method that will allow future generations to continue enjoying nature. Although we reconstructed it, this environment maintains the essence and characteristics of the original forest. Visitors to this place do more than simply enjoy it; they acquire a deep respect for life in their hearts and describe the experience as extraordinary. This conservation model shows surprising results. It opens up the possibility of counteracting the captivity of fauna and raising humanity's consciousness about the importance of protecting every species with respect and in total freedom.

The place has also become a living seed bank. It holds valuable forest and biological resources, together with cultural knowledge that provides extraordinary environmental services for life. Although I recognize the value of reforestation programs for conservation, this project goes further: it focuses on the holistic restoration of ecosystems.

I do not tell a tale of dates and unrelated facts; I narrate lived memories, witnessed events, and the human decisions that drove the forest's destruction. But I also offer the testimony of a passion born in my childhood. This passion led me to make the most difficult decision: to dedicate my life and my family's life to a mission many considered impossible. I made this choice knowing that, during the processes of colonization and civilization, society failed to correct various mistaken actions. Instead, people who think only of economic enrichment based on exploiting the forest's resources have strengthened these destructive practices, even to the present day. Their exploitation amounts to nothing more than the massive destruction of a natural paradise, and the resulting fallout continues to devastate the flora, fauna, and Indigenous nationalities.

This passion for saving nature brought countless insults, severe persecution, and criticism from figures who, in that era, lacked any vision of conservation. I did not waver in fulfilling my purpose; I simply waited with patience to demonstrate that "this sacrifice was worth it."

This research required decades of work before I saw results and gained enough certainty to share my positive experiences. I present these methods as the best—and sometimes the only—ways to solve the growing environmental impacts caused by humanity itself, driven by a desperation to find income and livelihoods based on exploiting everything that exists in nature.

WHO ARE THE ACTORS THAT SHOULD PROTECT THE JUNGLE?

These are the people or groups who, from birth, possess the gift of understanding nature's value. Alternatively, when they visit nature, it inspires them to change the course of their lives. They realize that every being on Earth possesses the right to live, and they recognize how important each individual is for maintaining the planet's balance.

We must understand and value every minuscule being that appears around us and every blade of grass that emerges in our path. Little by little, we must discover and ask ourselves: what is our mission in this world as human beings? We must then value nature and the actions we must undertake every day of our existence. If we understand the perfection of creation, we will realize that we were born with a mission: to return to the earth the very nature we stole from it during our lifetimes. Then we will become more sensitive to each living being and realize that we are not alone. Something divine manifests along the way; without our noticing, the restoration process will yield perfect results entirely without human intervention.

If you did not feel a passion for nature as a child, you can still cultivate that passion as an adult. Consequently, we must retrace a child's steps and begin observing nature as an ally in our lives. In this way, we will understand what our environment truly lacks, whether we live there or simply pass through. We will also comprehend the absolute necessity of lending our strength to the most fragile areas. Above all, we must save the species facing total extinction—species that would otherwise need hundreds or thousands of years to grow naturally again in their places of origin.

CHAPTER I: "MEMORIES AND LIVED EXPERIENCES"

Part 1. Extractivism: The Origin of the Destruction of Flora, Fauna, and Cultural Deterioration in the Ecuadorian Amazon

This chapter analyzes events that occurred in the Amazon. It shares my memories and observations, offers glimpses into my personal life, and presents my methodology for reconstructing the ecosystem alongside accounts from my research.

I experienced both errors and successes, but time and ongoing work continue to confirm extraordinary results. Therefore, I affirm: this story is true, the activities I undertook succeeded, and any effort to restore nature must inevitably include some of these actions. Practitioners can adapt or avoid certain elements depending on the implementation region, which will naturally yield different biological results.

Ever since I could reason, the mysteries of the rainforest fascinated me. I delighted in the birds, the forest, the animals, and the beautiful rivers flowing with crystalline, pure water. Many people told me that when you clear the forest, "nothing happens to the rainforest," and that "the trees will never run out because the forests regenerate on their own." But they were only justifying their actions. They had not yet felt the consequences of environmental devastation. As the years passed, rapid destruction advanced, threatening to extinguish essential elements of the rainforest, including human life.

The rainforest will weep, but its tears will flood its aggressors. Using its innate wisdom, it will restore itself, achieving perfection over thousands of years. Surely, humans will attack it again, as we struggle to maintain a high level of consciousness; we forget quickly and easily succumb to greed.

I want readers to imagine: What did the resplendent rainforest look like during the 1960s in the mysterious Ecuadorian Amazon? I was about four years old, but my mind has never forgotten; I could reason well enough to begin understanding the grandeur of life. Cities did not exist. I walked only through forests, surrounded by nature that no human being had ever destroyed. I considered it sacred.

The pure air emanated freshness. I heard birds singing constantly, while animals sporadically cried out or roared. As a small boy, I know I heard the jaguar; my uncle would say, "That's the panther!" The frogs joined in, creating a music that never disturbed the peace. Insects buzzed to signal summer, changing their melodies to announce approaching rain or nightfall. Abundant butterflies and insects gifted a diversity of colors to the environment as they flew. Nature produced music with a melody no one has ever imitated. The massive rivers roared loudly, and I could feel their fresh breeze from great distances; they carried absolutely no contamination. Even the smallest streams flowed with crystalline waters, where fish shined like mirrors, marking the finest expression of life. At the same time, the earth gifted its best products to feed the people who decided to colonize the rainforest. With absolute certainty, this paradise was perfect.

Undeniably, nature was the most generous living being. It gave so much and asked for nothing in return. I believe that, to live within it, we should have asked for permission and considered the rainforest a living ally in our existence. Destruction would cause it great pain. Respecting and defending it was the only rational choice; we should have planned to return everything we took from it. Humans made absolutely no effort to create it. Perhaps someone planted it thousands of years ago to ensure its existence; therefore, we had a duty to honor that legacy of conservation. We also had to recognize the divine action behind its perfection, an order that made our lives possible. Regardless of your beliefs, nature unquestionably belonged to the sacred.

I was only four years old. Naturally, I did not perfectly understand my parents' mission. Perhaps they yearned to live in the rainforest. Later, I realized that, like many colonizers, they sought a new way of life. Today, I feel I am reliving that childhood. Our family included my parents and three brothers; I was the middle child. We lived in the Colonia Azuay, in the Topo settlement, the gateway to the Ecuadorian Amazon. About twenty years ago, I returned to that place. I felt such profound nostalgia and pain! Nothing remained of the nature that I still carry etched in my childhood memories.

Part 2. Causes That Contributed to the Threatened Extinction of Flora

Destruction of Ecosystems

As a child, I viewed the world with an innocent consciousness and gaze, but I still remember the loud sound the great, majestic trees made as they fell. Back then, colonizers cut them down using axes. As new inhabitants arrived at the entrance to the Amazon, I heard axes striking the great trees wherever I walked, until the massive trunks crashed to the ground with powerful impacts. Falling trees tore apart the surrounding vegetation. I also frequently heard people shouting warnings: "Run and protect yourselves so the trees don't crush you!" I can still recall the characteristic morning sound of men sharpening their machetes on special river stones.

At that time, the colonizer faced one primary challenge: destroying the forest to plant crops. People used very little wood. Men operated two-person hand saws to cut beams and planks to build their own houses on-site, or to sell in the newly developing settlements. Many of these trees housed nests of birds and small animals, which died when the trunks crashed down. Surviving animals fled, frightened or dying from their injuries. Meanwhile, birds cried out in search of their agonizing young, never imagining they had lost them forever (Figure 1).

Figure 1 - Deforestation affecting birds and animals

In the streams or springs near their houses, people placed pieces of bamboo to collect water for cooking, bathing, or washing clothes. I remember the crystalline, cold, and clean water. We drank directly from the spring or from a bamboo spout that we positioned at the perfect height for our needs.

Sadness and melancholy invade me. I have never again stood so close to paradise. The memory feels like an unforgettable dream because the destruction struck so violently.

At six years of age, my life changed completely. I began preparing to attend school in the Topo settlement. My parents had already taught me to read and write. Later, we moved to the Río Negro Parish, and I finished my schooling in the Mera Canton at the Fray Jacinto Dávila School. I clearly remember my teacher telling us: "Children, people will soon extinguish certain trees." He also taught us: "We live on Mother Earth, and we must care for her and respect her." During this period, my brother Efraín and I spent our free moments walking to nearby rivers or forests. From a very young age, nature's energy imprinted itself upon my heart. I set a lifelong goal: to live in a rainforest and, if the Divine Being allows it, to die there.

Even in that short time, I found it difficult to locate a nearby rainforest like the one that had surrounded us just a few years earlier. The governing system forced us into a new modality of existence. It tried to hide its dark purpose behind deception, falsely promising to improve the lives of all inhabitants in the Ecuadorian Amazon.

Colonization of the Ecuadorian Amazon

I was about eight years old. I remember people spreading rumors: the government was pushing the colonization of the Amazon with greater intensity. Consequently, the towns and peasant farmer populations grew. These new arrivals came primarily from Ecuador's Andean region. Families began replacing forested lands with small farms and large pastures to survive.

During this era, the Ecuadorian State sought to seduce many families by advertising the benefits of colonization. To achieve this, the government issued the country's First Law of Agrarian Reform in 1964. It strengthened this effort by ratifying the law in 1973 and creating the Ecuadorian Institute of Agrarian Reform and

Colonization (IERAC). This institution took charge of adjudicating—or giving away—fifty hectares of virgin rainforest to any family wishing to join the supposedly attractive task of colonizing the Ecuadorian Amazon (Figure 2).

Figure 2 - IERAC Seal

(Image recovered from the internet and clarified with Nano Banana Pro CC-BY-4.0)

Consequently, many families arrived from other regions, expecting to improve their standard of living and aspiring to a promising future. State policies framed this migration, however, to facilitate petroleum extraction. The government progressively justified this exploitation by arguing that timber and fauna resources would never run out because they were renewable.

With apparent success, the IERAC granted untouched rainforest lands for colonization. Surely, no one realized that this law and the accompanying forestry exploitation policies were launching a massive slaughter of the planet's most important environmental fortress. This destruction would eventually trigger global repercussions, as the Amazon Rainforest eternally served as the lungs of the world.

The government distributed these lands with provisional documents and agreements, requiring the colonizing family to clear at least eighty percent of the forest (Figure 3). IERAC officials would then conduct an inspection to verify compliance. Once satisfied, they ratified the family's right to absolute ownership of the adjudicated lands and granted them the public deed.

Figure 3 - Deforested land

Consequently, the first colonizers massively developed pastures for cattle ranching, planting primarily gramalote grass (Figure 4). They cleared thousands of hectares of forest for this purpose, completely ignoring the distance to access roads. They left the dead wood of millions of fragmented trees unused; they cut down these ancient forests solely to cultivate vast pastures and legalize their property ownership (Figure 5). I remember people calling the forests "idle lands" because they produced nothing; this label made clearing and cultivating the land an obligation.

Figure 4 - Gramalote (Axonopus scoparius)

Figure 5 – Amazon land cleared for cattle pasture

This labor demanded great sacrifices from the colonizers, who continued using simple tools like machetes, axes, and hand saws. Furthermore, they lacked knowledge about the existing timber species. An exhausted farmer would finish an arduous workday, arrive home, and comment during his rest: "Today I only managed to cut two or three trees." He never imagined, of course, that he had felled trees that had lived for hundreds of years—

incorruptible, fine woods whose dense nature made them incredibly rigid and difficult to cut.

Figure 6 - Mountain cabbage (Vasconcellea microcarpa)

Bridle paths served as the only communication routes to reach a town, suitable only for traveling on foot, on horseback, or with mules. I also remember colonizers using oxen, finding them more agile for navigating paths filled with deep water and mud holes. I do not recall any markets existing; small shops operated in the towns to ensure the people's survival, though families depended largely on their own crops. I remember a wild vegetable people called mountain cabbage (Figure 6). My mother frequently sent me to gather this plant's leaves so she could prepare our meals.

How could I forget how much sacrifice the peasant farmer's work demanded? The lands took a long time to produce any economic return, and the meager profits barely covered daily needs. These families never gathered the fortunes they so deeply yearned for.

As the years passed, numerous peasant farmers found themselves unable to continue their arduous fieldwork. After fulfilling the regulatory time IERAC imposed to secure possession of their territories, many began selling their properties. They sold to neighbors or to newcomers from other regions joining the cattle ranching boom. These buyers accumulated massive tracts of land, eventually establishing themselves as large landowners with estates spanning hundreds of hectares.

Published in the newspaper El Diario. August 2010. www.eldiario.com.ec

> The seizure of political power in Ecuador by a military dictatorship (1963), which brings into effect the first Law of Agrarian Reform in the country. On July 11, 1964, through Decree 1480, the Military Junta of the Government issues the Law of Agrarian Reform and Colonization.

Law of Agrarian Reform (1973).

> On October 9, 1973, the Military Junta issued Decree 1172 (published in Official Register No. 410 on October 15, 1973) with the name Law of Agrarian Reform.

The now-defunct Ecuadorian Institute of Agrarian Reform and Colonization (IERAC) dedicated itself to the thankless task of distributing massive state territories—virgin lands abundant in flora, fauna, and water sources—to contemporary politicians and to active and retired military personnel. The new owners quickly cleared these forests to exploit the wood and convert the terrain into pastures and short-cycle crops. Shortly after, these destructive practices turned many areas into wastelands and devastated their watersheds.

Ironically, the former IERAC became the greatest culprit for driving deforestation and fracturing the landscape into tiny, unsustainable farms in Ecuador. The agency drove the colonization of uncultivated lands using technically unsound methods. Officials failed to balance biodiversity with technical agricultural practices; they ignored the need for proper watershed siphon systems and failed to require confined cattle management.

> During the era of the military triumvirate and the government of Guillermo Rodríguez Lara, state policies actively forced farmers to deforest the mountains. If farmers left the trees standing, IERAC would classify the properties as idle lands and intervene to confiscate them.
>
> The provinces of Manabí and Esmeraldas suffered some of the worst impacts. IERAC's policies flooded these regions with colonists, especially from Pichincha—primarily hacienda workers whose employers generally lived in Quito. These newly arrived colonists established towns such as La Concordia in Esmeraldas and La Manga del Cura in Manabí, carving out territories that today look like no-man's-lands.

Traces Left by Monocultures.

The apparent success of cattle production drove colonists to intensify their pasture cultivation. While the Amazon does possess prodigious regions where high-quality pasture has endured for years, people have used this fact merely to justify the massive clearing of nature. In reality, the majority of pastures suffer from poor conditions and depleted nutrients. A single cow requires more than a hectare of pasture to survive, forcing ranchers to provide supplementary feed just to maintain efficient cattle production.

This monoculture has completely eroded the soils in several sectors, driving up the economic costs for anyone attempting agricultural or forestry activities on these lands. Consequently, the environmental impact remains irreversible in the short and medium term. Cattle pastures became the Amazon's first monocultures, representing the first harsh lesson nature taught humanity—a lesson we have yet to assimilate.

I also remember how the cultivation of the naranjilla fruit (*Solanum quitoense*) became predominant among the colonists during that same era (Figure 7). Commercializing this fruit provided farmers with enough economic return for their families to survive. However, the intermediaries who transported the fruit to city markets across the country reaped the greatest economic advantages. Consumer markets widely accepted this produce, dubbing it the "golden seed" for its exquisite flavor and nutritional value. This high demand drove up production, generating attractive profits for both the peasant farmer and the merchant.

Figure 7 - Naranjilla (Solanum quitoense)

To supply the market, farmers fragmented exuberant virgin forests. Initially, they cut down nearby woods, but they increasingly expanded into distant territories, often walking up to a full day just to reach their cultivation sites.

The farmers carried naranjilla seeds on their long journeys. First, they cleared a small patch of vegetation to sow the seeds. Next, they slashed the low undergrowth across many hectares and immediately felled the towering trees. They then performed the repique—cutting the branches off the fallen trees to make room for the crops. By this time, the seedlings in their makeshift nurseries were ready for transplanting.

After transplanting the crops, some peasant farmers returned home to tend to other tasks. They would come back approximately two months later to weed—slashing away anything they considered a nuisance—and to replace any dead seedlings. From then on, they made periodic visits to keep the crop weed-free.

After about ten months, they hired workers and horses to harvest the fruit and haul it to collection points where the farmer sold it. Workers packed the fruit into wooden boxes, sorting it by size—large, medium, and small—and designating it as first, second, or third-grade naranjilla. Finally, the buyer transported the boxes via trucks to other regions of the country, selling the harvest directly to consumers in markets or in bulk to other wholesalers.

This massive destruction of nature and intensive cultivation caused an irreversible impact. During this period, when a farmer hired workers and horses for the harvest, he often arrived to find fallen, damaged fruits and yellowish plants ravaged by pests and diseases. This devastation occurred simultaneously across all regions where farmers cultivated this fruit. It created a problem that growers have failed to evade for over sixty years, making it impossible to cultivate this species organically today. In most sectors, even chemical applications fail to save the crop. Consequently, markets today only sell hybrid naranjilla fruits grown with heavy chemical applications.

Surely, this served as the second lesson nature taught humanity. The massive destruction of the forest to support naranjilla monocultures triggered another irreversible environmental and social impact. Intensive monocultures simply cannot survive in the Amazon; nature always intervenes to regulate the imbalance. Yet, humanity refuses to learn this lesson. Farmers continue planting intensive monocultures. When problems arise, they address them by applying chemicals, which only multiplies the environmental, biological, and social damage.

Because society never records these stories, farmers repeat the same errors. They plant massive monocultures, either on soils already destroyed by cattle ranching or by cutting down thousands of hectares of primary forest and totally eliminating the native vegetation. Consequently, insects, wild birds, and animals lose their habitats and food sources. Forced to change their eating habits, these creatures transform into pests that consume the plants, flowers, and fruits of the monocultures and wild seeds. To counteract them, growers increasingly resort to chemicals. This triggers yet another environmental and social disaster: the massive death of insects, spiders, bees, and other crucial species. Growers intentionally apply specific poisons to murder each of these living beings that are so valuable to life.

Currently, diverse impacts continue to increase. In the Ecuadorian Amazon, developers constantly expand land subdivision and commercialization programs to grow urban areas across many kilometers. These new subdivisions eventually merge with the cities, generating threatening impacts by contaminating springs, streams, and rivers. Thus, these development activities multiply the environmental and social damage.

We continue to lose natural spaces. Because many of us enjoy the privilege of living in the Amazon, native nature should surround both large and small populations. City planners could design majestic, educational parks that feature environmental and cultural interpretations, deliberately minimizing the use of cement and artificial materials.

As a food staple since the era of colonization, Indigenous people and farmers have cultivated plantain (*Musa spp.*), yuca (*Manihot spp.*), papa china (*Xanthosoma spp.*), and chontaduro (*Bactris gasipaes*). Initially, these organic agricultural efforts produced extraordinary yields. Growers even thought they could sell their produce to populations in the Andean region. However, land erosion quickly undermined this activity. Currently, farmers maintain these crops only in defined zones featuring better soil conditions. They also clear primary forests to secure excellent harvests, which supply their families and leave a surplus for commercialization. Because of intense rains, the harsh climate, and fragile soils severely limit the species farmers can cultivate in the Amazon, inhabitants in several local towns now depend heavily on food produced in the Andean and Coastal regions.

Farmers must balance the cultivation of any product in the Amazon with an adequate proportion of native nature. This combination maintains equilibrium. While this approach requires more territory, planting or preserving tree barriers and nature reserves next to crops ensure that insects, birds, and animals retain their natural food sources. Consequently, they do not attack the crops. We must understand that repairing heavily destroyed zones requires significantly more time to reach this balance. Farmers must diversify species in their crops to conserve a biological and food equilibrium. If we put this suggestion into practice, we will successfully achieve environmental conservation and restore the organisms that constitute the food chain. When we destroy nature, insects, birds, and animals change their eating habits, and our actions transform them into pests. Out of ignorance, most people try to secure their production and investments by resorting to intensive chemical use as their only alternative. This inevitably creates irreversible environmental and social impacts.

Replacing diseased or pest-ridden plants in a monoculture with other plants of the same family—even resistant varieties—does not solve the problem. In a short time, the same issues will afflict the new plants, proving that we have learned nothing. Instead, farmers should establish any new cultivation using an Agroforestry approach.

Intensive Commerce of Timber Trees and Palms.

To secure short-term economic livelihoods and benefit the towns, colonists, and Indigenous communities, colonizers intensified the exploitation of nature's products. To achieve this, the government expanded road construction, arguing that these routes would help communities transport their agricultural harvests to markets more successfully and supply populations that desperately needed those products.

Thus, the opening of new motorable roads led to a progressive discovery: despite the massive environmental destruction caused by colonization policies, fine timber still existed in the Amazonian forests. Therefore, politicians and loggers heavily prioritized timber exploitation. They erroneously considered vegetation a naturally renewable product, believing that if they cut down trees and palms, many more would simply sprout and grow naturally. Furthermore, they argued this logging would provide a permanent, sustainable business.

This mindset established a profoundly misguided livelihood. By then, chainsaws were already roaring deep inside the majestic forests, completely ignoring the distance from the new roads (Figure 8). Loggers cleared trails for horses and, in some cases, for tractors, which only accelerated the destructive activity. The city of Puyo, for example, hosted numerous collection centers, countless factories for preparing and commercializing wood, and creative artisans who crafted eye-catching furniture.

Figure 8 - Felled tree

Builders easily obtained wood to construct houses, finding species that perfectly suited the taste or greed of each person (Figure 9). Buyers passionately sought out specific colors, durabilities, and smells, selecting from species including (species in table 1):

COMMON NAME	***FAMILY***	***SCIENTIFIC NAME***
Caoba	FABACEAE	*Platymiscium pinnatum*
Guayacán	BIGNONIACEAE	*Handroanthus chrysanthus*
Bálsamo	FABACEAE	*Myroxylon balsamum*
Pilche o Chanúl	HUMIRIACEAE	*Humiriastrum procerum*
Guambula	OLACACEAE	*Minquartia guianensis*
Ahuano	MELIACEAE	*Swietenia macrophylla*
Moral fino o bobo	MORACEAE	*Clarisia racemosa*
Laurel	BORAGINACEAE	*Cordia alliodora*
Cedro	MELIACEAE	*Cedrela odorata*
Chuncho	FABACEAE	*Cedrelinga cateniformis*
Chonta	ARECACEAE	*Iriartea deltoidea*

Table 1

Figure 9 - Cut trees

For a long time, nothing restricted the clearing of forests. The peasant farmer determined the quantity of trees cut and the quality of the wood based on market demand. Eventually, a State Institution intervened to regulate this free trade, becoming the sole provider of timber exploitation permits. These authorizations exceeded 250,000 hectares of primary forests each year.

Consequently, this activity has increasingly generated incalculable environmental impacts. It clearly demonstrated the deadly destruction caused by wood extraction, driving the disappearance of flora and fauna in the rainforest. Without a doubt, some insects, birds, and animals became extinct before we even knew they existed.

We must not forget that the peasant farmer's hope and illusion of obtaining a fortune ultimately consumed his own existence. Cutting trees, making planks or beams, and carrying them on his shoulder or on horseback along inhospitable, distant paths in sun or rain just to sell the wood by the roadside was purely a desperate struggle for survival. The money earned barely covered daily expenses, supported his family with severe limitations, and paid the installments on the chainsaw he had bought on credit with great difficulty. I do not remember, nor do I know of, a single peasant farmer who made a fortune or lived comfortably from this activity. The truth is, every peasant farmer sacrificed his youth, his life, and his family's well-being, always hoping for the better days promised during the colonization processes. Many died waiting. True fortunes generally appeared only when heirs, uninterested in the land because they had other livelihoods, sold the parcels entirely.

The people who paid the peasant farmer a minimum amount for the wood left by the roadside were the ones who truly capitalized on the wood's authentic value and the economic fortune nature produced. Those who owned factories to process and resell the wood at a higher price amassed a fortune, as did those who created large export companies, as the finest woods were already highly coveted globally.

This inequality of opportunity left a tragic legacy for nature, and equally tragic consequences for the people who initially trusted the promised opportunity to make a fortune from the sacred rainforest. These people received no advice or support to sell the wood at fair prices, nor were they warned about the devastating consequences logging would inflict on their lives and on fragile ecosystems.

Sporadically, people tried to revive the economic activity by reforesting these timber species. This proved impossible because the coveted fine woods take many generations to grow, mature, and develop their characteristic quality, durability, and coloring. Furthermore, these efforts completely ignored environmental conservation.

Many people had the opportunity to build their houses, furniture, doors, or windows using wood from centuries-old or millennia-old trees. These incorruptible woods last for generations and must be conserved today because society will not achieve such luxury again. One can only conclude that loggers murdered these trees, sacrificing their lives and the lives of all the beings that inhabited them to satisfy human greed.

Currently, the majority of fine wood species are threatened or in danger of extinction. Seeing even a small tree of certain species has become a luxury for society. It is no longer accessible even for local inhabitants to embrace, know, and enjoy these valuable specimens. Only people who can pay fortunes to travel deep into the Amazon interior might sporadically encounter these slow-growing trees, realizing that tales of their past abundance sound like legends. Furthermore, commercializing any rainforest product without prior planning inevitably leads down the path to extinction.

Intensive Extraction of Orchids.

The year is 1969, and I am 13 years old. My parents, my brothers, and I move from the Mera Canton to the city of Puyo, fifteen kilometers further into the Amazon. I remember arriving in an extremely small town. I quickly explored the nearest rivers: Río Puyo, Pindo Grande, Pindo Chico, and Pambay. They were magnificent. All of them boasted pure, crystalline waters rich in fauna. Large fish inhabited these waters, feeding the community and attracting the few local inhabitants. People could easily navigate large stretches of these rivers; we frequently watched inhabitants of various Indigenous communities guiding their canoes. We, along with visitors from the Andean region, enjoyed the beautiful natural swimming spots.

Despite the approaching catastrophe, we still found abundant vegetation to admire. Diverse species of heliconias, bromeliads, and orchids provided vivid attractions and flashes of color within nature's dominant green. Hummingbirds and butterflies competed to drink the flowers' exquisite nectar. Nature still maintained diverse landscapes that emanated peace and hope; it remained part of a paradise that, unfortunately, would not last much longer.

I watched large groups of foreigners annihilate these beautiful places. These collectors possessed sophisticated equipment, specialized knowledge, and official authorizations issued by Ecuadorian State control agencies. Therefore, they could freely collect whichever wild plants interested them from any location they chose. They targeted primarily orchids—flowers that to this day inspire tenderness and admiration but also provoke an increasing greed to possess them.

Figure 10 - Orchids growing from trees

Depending on the species, these plants grew on towering trees, on rocks, or on the ground. The collectors extracted them using long poles or specialized equipment. At the end of the day, they cleaned, washed, and packed the plants into massive cardboard boxes. Historically, these orchids had endured for thousands of years, living freely in the rainforest. At that moment, however, they became prisoners crying out for freedom, but no one heard their cries.

In a short time, collectors shipped them far away, subjecting them to controlled climates and habitats and inevitably cultivating them with chemicals. Like any living beings facing harsh conditions, some resisted, while others died. These extracted plants left behind their pollinating insects—creatures that depended entirely on the perfume of the orchid flowers to reproduce. Spiders and other beings that required these specific plants to survive either died or had to drastically adapt their ways of life because the orchids had vanished or grown scarce. Today, scientists have assigned specific names to these missing plants, progressively archiving them in library tomes and adding them to the growing lists of endangered species.

Humans continually launched an onslaught against nature—an assault that surpasses human understanding. People armed themselves with chainsaws, tractors, machetes, weapons, and destructive equipment, producing massive contamination. Nature, however, armed itself with the wind, the rain, its rivers, its shade, its fauna, and its enduring mysteries. Surely, nature will use these remaining forces to defend itself whenever it can no longer bear human aggression, just as it continues to do today.

Society completely lacked knowledge about environmental impacts, perhaps blinded by the sheer abundance of past eras and nature's boundless generosity. Almost no one realized the ecological disaster these activities would cause. Only a few years after loggers and farmers destroyed the original forests, vast territorial extensions gave way entirely to monocultures. Peasant farmers found it increasingly difficult to survive on these large properties because nature had vanished, the soils had eroded, and contamination had ruined the waters. With some luck, certain territories naturally regenerated into secondary forests. However, these newer forests support only limited plant and animal species; the original species have disappeared completely.

Part 3. Causes That Contributed to the Threatened Extinction of Fauna.

Wildlife Trade.

The destruction of ecosystems at every stage of development obliterated the habitats of all living beings in the intervened zones. This devastation drove many species to total extinction—species we surely never had the fortune to know.

During the 1960s, society actively began shattering the lives and freedom of the rainforest's inhabitants. I remember people from other cities—including inhabitants of the Baños Canton—who made their living by trafficking fauna. They surely believed this resource would never run out, given the immense numbers of bird, animal, and insect species present in that era.

I was a child, but I remember perfectly sitting in our small yard or on the porch of our humble country house in Colonia Azuay, above the Topo settlement. Suddenly, strangers would arrive, often accompanied by foreigners. They would scout for the most suitable locations and string large mist nets and traps between the trees. In a short time, the nets would begin to shake. You could see the birds' desperate struggle to free themselves and resume their normal flight, but it was already too late; they were completely entangled. The more they fought to escape, the more entangled they became. Some died from sheer exhaustion or suffocation.

These trappers managed to capture hundreds of birds. They selected those with the brightest plumage and the most perfect, symmetrical natural designs characteristic of these species. Then, they stuffed them into cages. No matter how violently the birds beat against the bars, or how much anguish they felt, they were trapped! The birds knew intuitively that they were leaving behind their nests and condemning their chicks to die. Yet, while the birds reflected pure anguish and desperation, the hunter felt only excitement, seeing dollar signs for each capture.

Figure 11 - Birds caught in a mist net

Many birds died before reaching their destinations, but the traffickers still sold their feathers. I remember seeing them pack hummingbirds in cloth wraps like belts, stacking them on top of each other in boxes with their beaks pointing in the same direction. I also remember that some birds died the instant human hands touched them, suffering heart attacks from the sheer terror.

They hunted the most beautiful butterflies for the exact same purpose. Collectors caught some with their bare hands and chased others with nets (*jamas*) until capturing them. I remember watching them pack the insects into paper envelopes and place them in boxes. Naturally, this business generated significant wealth for the collectors during that era. Society lacked any consciousness about the sacrifice that captivity and death represented. Furthermore, this trade operated legally and without restrictions; the hunters even held official authorizations from the Ecuadorian State institutions regulating these activities at the time.

Thus, they freely extracted whatever fauna they desired and exported it wherever it proved most convenient. In a short time, these birds joined the growing lists of endangered species.

I also recall teams of hunters arming themselves with sophisticated weapons to hunt nature's most majestic beings:

jaguars, bears, deer, and more. When they successfully killed a specimen, society celebrated these hunters as remarkably brave men. Upon returning to their towns, they received decorations, trophies, and prizes—though I no longer remember exactly who awarded them. With great pride, these hunters exhibited their sophisticated weapons, pelts, mounted animal heads, and the story of each kill in their private museums. It was a profoundly unequal fight. An animal or bird could not defend itself; they possessed only their beaks or claws, while the hunter faced no risk of even getting close, firing weapons from great distances.

During this same period, the commerce of various tamed wild animals and birds as pets proliferated across many towns in the Ecuadorian Amazon. The towns remained small, and their inhabitants submissive. I remember several foreigners who would set up on a house porch or in the town square, sitting in chairs beside a table. They hung signs written in very clear letters that read: "we buy frogs," "we buy butterflies," "we buy beetles," "we buy lizards," or "we buy snakes." Unaware of the future consequences, locals dedicated themselves to hunting these specimens to earn a few cents of *sucres* (the Ecuadorian currency at the time) from these foreigners. However, depending on the importance the buyer placed on each species, I watched them inject many of these creatures and place them inside containers filled with chemicals to prevent their deterioration.

Captivity of Fauna.

People commonly found cages holding birds or animals in town houses or in the countryside. Depending on the species, owners also tied them near the foundations of their homes, in their patios, or to a tree trunk. Similarly, captors clipped the wing feathers of some bird species so they could not fly or escape. People generally practiced this on parrots, guans, toucans, and other animals. This widespread practice and greed for keeping tamed wild animals spread throughout the country. I remember some people saying, "I came from Quito and I want to take one back," or "Someone asked me to bring back a monkey, a parrot, a macaw," etc. They went to the market or visited a house and bought the animals or birds they wanted.

Buyers and sellers drove this trade without considering the pain captivity caused. On the contrary, humanity still has not learned the lessons nature teaches. People do not feel the pain of others, or they believe animals do not feel pain; yet people speak of freedom, and governments even pass animal rights laws.

Allowing captivity equals loving prison.

Currently, society practices captivity as the only alternative to protect wildlife. Mud-filled cages and pens evidence this mistaken practice. Zoos proliferate in the Ecuadorian Amazon, expanding these eternal wildlife prisons that masquerade as nature. Captors ignore the sadness in the birds' and animals' eyes, and their desperation to escape. Observers can see the stress these animals feel in their behavior: they no longer move continuously from one place to another. They manifest their sadness when they pace back and forth in the same spot inside their cage or pen, because they have nowhere else to go. Consequently, their beauty evaporates as their hope for freedom vanishes. Anyone can clearly notice this decline in the quality of their feathers or fur, because they no longer shine as they did in the wild. They can no longer eat their chosen diet; now, in prison, they must satisfy their hunger with foods they perhaps never tasted before, because those foods did not exist in the jungle. Their blameless lives and their hope for freedom vanish until death finds them in a prison that never stays empty, because another victim of human greed will soon occupy it.

Captivity does not symbolize protection, and dependence does not mean freedom.

Part 4. Causes of Cultural Deterioration in the Amazon.

Characteristics of the Different Indigenous Nationalities of the Ecuadorian Amazon and the Disruption of Ancestral Harmony

Regarding the Indigenous nationalities that inhabit the Ecuadorian Amazon, we must remember the close relationship they maintained—and some still maintain—with the ecosystems since ancient times. However, the incessant destruction of the jungle, the pressure from oil and mining companies, the lack of water due to river pollution, and harsh government policies actively change and deteriorate much of their ways of life. Furthermore, the progressive disappearance of foods that nature naturally provided them since remote times adds to this problem.

Therefore, society must know that the flora, fauna, and Indigenous nationalities forever shared a close link. They represented elements that never separated. Together, they constituted the food chain. Consequently, a perfect equilibrium existed within the ecosystems. The deceptive method of promoting civilization and colonization deteriorated this harmony, producing a shocking impact on each of the different nationalities. This process forced changes upon their ways of life, constantly causing deterioration while only holding distant intentions of improvement. Now, I do not believe the actors of destruction acted out of ignorance. Governments exert a permanent pressure against the communities living inside the jungle, and this pressure expands very clearly, always reflecting the achievement of grim purposes as their primary goal.

The False Promises of Civilization

The designers of civilization advanced their plans using small gifts and false promises, creating a profound submission sustained by hope. This caused more people to leave their jungle on their own, while family or friends brought others to the towns so they could observe the lifestyle in these progressively expanding settlements. Several of them wanted to leave their jungle to forget that they had existed within sacred and majestic nature for thousands of years. They failed to realize that nature gave them life, clothing, food, medicine, weapons, housing, and everything they needed, forming within them a true science and the most sublime strength, which we summarize as Ancestral Knowledge.

Without realizing it, foreigners stripped them of this knowledge. With little more than a dollar or a few small gifts, outsiders tore this wisdom away from them for commercial purposes, before the Indigenous people could realize that their knowledge constituted the greatest strength of all Indigenous cultures. These deceitful actions also formed a core part of extractivism.

We must recognize that the wisdom regarding the properties and uses of the diverse Amazonian flora resided within Indigenous cultures. Society and the State never recognized that an ancient science existed within the elders of each tribe, as these elders discovered the properties of each plant and how people should apply or use it. For various reasons, they respected specific plants because they considered them sacred, always remembering that the different Indigenous nationalities of the Amazon represented just another component of the region's ecosystem. Therefore, ancestrally, they did not destroy nature. They used everything the jungle provided them according to the needs of their family or community. Excessive use did not proliferate because they did not commercialize their products and did not need to do so.

Establishing civilization without planning brought major consequences at different stages. Outsiders taught them new ways of living, but never taught them how to maintain their habitats or transmit their valuable customs and ancient knowledge to their children. Since then, most Indigenous inhabitants who maintain contact with cities seek ways to survive, often through activities far removed from their ancestral ways of life.

The Arrival of Commercial Exploitation

While Indigenous communities accepted this process of integration into the new lifestyle, people from other regions introduced new ways to make a fortune. These merchants took advantage of the situation to increase their wealth. Acting as great explorers, they entered the different communities, using the majestic rivers as transport routes, and generally traveled in the canoes of the Indigenous people themselves. These traders brought items that held little to no economic value, such as tin jewelry, combs, mirrors, or any object that caught the attention of the jungle inhabitants. They exchanged these items for pets or animal skins—from jaguars, bears, deer, boas, etc.—skins that the Indigenous people used as blankets or clothing. As a consequence, this unfair trade continued until the traders almost completely stripped the different communities of their animal skins, committing yet another act of extractivism.

The greed of merchants to make substantial fortunes with this lucrative business drove them to find another strategy to facilitate hunting. They introduced rubber shoes or boots, machetes, knives, firearms, gunpowder, and ammunition so the Indigenous people could hunt the animals and obtain this merchandise more easily. Buying skins for ridiculous prices and reselling them at high prices allowed the merchants to obtain huge economic benefits. Thus, submissive and excited, the Indigenous people considered these items incredibly precious. They no longer had to set traps or track animals to kill them with spears; they just fired a shot and easily hunted whatever they wanted. In a short time, the different communities ran out of food.

During this process, another survival alternative emerged for the Indigenous people: merchants created markets to sell birds and tamed animals as pets, including macaws, various species of parrots, monkeys, guans, toucans, ocelots, armadillos, otters, and more.

I remember that, at that time, people identified the Ecuadorian Amazon as *El Oriente Ecuatoriano* (The Ecuadorian East). People from the Andean region very commonly told their friends or relatives, "Are you going to the Oriente? Buy me a monkey or a parrot," or whatever animal or bird they wanted. The price increased if the parrot knew how to talk. Then, upon returning to their city, the traveler fulfilled the request by saying, "Here is your macaw," or whatever other animal the person had asked them to buy.

The Global Trade of Ancestral Heritage

At the same time, traders in the towns opened markets for various adornments made with the bones and feathers of the most beautiful birds. The Indigenous people ancestrally used these objects for personal wear according to their hierarchy, or for parties, rituals, or home decorations, and their practices never harmed nature. However, this growing trade incentivized the Indigenous people to hunt more wildlife to meet the demands of the merchants.

The buyers of these products resold them to others at very high prices, and even exported them to other countries. In their destination countries, buyers considered this fauna and these crafts high-value relics, purchasing them to display in their homes as souvenirs or exhibitions. Therefore, museums in other countries now hold this great history and ceremonial objects, because people assigned them little value in their places of origin, where the payment for them remained virtually imperceptible. In this way, the impacts on wildlife and on people of Indigenous nationalities increased. No one realized the situation or intervened to prevent this immense tragedy from unfolding in a short time. Now we must remember history and ask ourselves: Who made a fortune with these invaluable ornaments? Certainly, the Indigenous people did not, as they did not receive even a fraction of that fortune.

People also routinely found shops in the towns selling live or dead (embalmed) wildlife and countless garments. Indigenous people supplied these shops while receiving minimal payments. Foreigners generally visited these shops, buying in bulk to resell in their home countries or to use as gifts, souvenirs, and evidence of their trip.

Infrastructure and Destruction

By 1980, Provincial Councils continued their development plans. They increased the opening of vehicular access roads to other communities in the Amazon, connecting them with river transport. This river transport served as a means of communication for the different communities, who perpetually traveled along the majestic rivers, using large or small canoes according to their needs. They built these boats primarily using Cedar (*Cedrela odorata*), Ahuano (*Swietenia macrophylla*), and Chuncho (*Cedrelinga cateniformis*) trees.

Consequently, the government built the new roads to facilitate the Indigenous people's trade of food products to the towns, mainly plantains, cassava, taro, peach palm, and barbasco (a root used to catch fish). They cultivated these products on their farms. Likewise, this connectivity allowed the inhabitants of Indigenous communities to stock up on products sold in town stores. They primarily bought salt, sugar, shotguns, ammunition, and gunpowder for hunting. They paid for these goods with garments, ceramics, pets, and I also remember them paying with gold.

To construct the highways, workers cut down all types of vegetation, including centuries-old and millennia-old trees. The workers buried these trees in swamps to facilitate the construction of the new roads. This destruction obstructed animal trails, killed birds in their nests, and eliminated shelters or burrows. Drivers and workers hunted every animal that appeared on the road or intentionally chased them down to run them over with their vehicles. They waged a war against everything that lived in the jungle. At the same time, loggers cut down the forests that remained alongside the highways, either to sell the wood or to grow grass for cattle ranching. Therefore, no one felt remorse for the destruction. People also seemed to think that animals or birds did not feel pain. On the contrary, the population felt a generalized excitement about finding an immediate economic solution—one that ultimately did not last.

This free trade persisted for over fifty years. No individuals or institutions stepped in to stop this ecological disaster. This failure caused a massive imbalance in nature and life in general, worsening the deterioration of Indigenous populations' lives and threatening the existence of these valuable Nationalities, as well as the flora and fauna of the Amazon.

Rumors also circulated suggesting that the government should relocate certain communities to other places that lacked oil. Anyone who understands the root of their existence knows that this would surely result in mass suicide. The lives of these communities still depend on the jungle. Most importantly, their ancestors chose their living locations based on their cultures. In these places, they find and maintain their plants, forests, animals, and sacred sites, all of which form part of their worldview. Without a doubt, their science disappears outside their natural environment.

The Ongoing Environmental Tragedy

When we understand that the Indigenous nationalities formed a component of the food chain, we realize that they did not merely live off the jungle or exploit it. Instead, they coexisted with the jungle in complete harmony.

I firmly believe that organizations should place the different Indigenous nationalities on threatened species lists, because society commits aggression against them by polluting their rivers. They used to fish, bathe, and navigate in these rivers, which now run thick with oil. In many cases, rivers act as literal drains for sewage, transporting all the pollution produced by various cities. Meanwhile, public institutions and society fail to realize the massive debt they owe to these communities. Because these groups historically settled primarily on riverbanks, many still wait for the false progress promised by the roots of civilization. Outsiders told them, "You must live with hope," but hope does not improve life. Fulfilling the promises made would be the logical action. They were promised hope as children; now they die as elders without ever receiving a single crumb of what society offered them.

We can witness a small example of the reality concerning the vast debt society owes to Indigenous communities in the Pastaza Province. Let us highlight two towns, Puyo and Shell, which remained small 60 years ago. The Puyo River borders the city of Puyo to the North, and the Pindo River borders it to the South. Both of these rivers possess many tributaries. Progressively, along the route between these two towns and at their edges, many other settlements have formed over several kilometers. These populations have grown exorbitantly, generating massive pollution. The companies and the cities themselves produce this filth and contamination, discharging all their waste into the tributaries and rivers.

I remember that at the mouth of the Alpayaco River flowing into the Pastaza River, as well as at the current bridge crossing the Pindo River toward the Puyo Hospital, and behind the Specialized Institute on the Puyo River, trucks daily dumped tons of city garbage. The dumpers then waited for a heavy river swell to wash the garbage away. Today, the city no longer dumps garbage in that manner; however, a garbage dump located just a few kilometers from Puyo continues to pollute a river.

Most terribly, countless Indigenous communities had historically settled along the paths of these rivers. These inhabitants fished, navigated, bathed, and drank from these pure and clean waters. They continue to do so today because they have no other alternative.

Authorities never worried about this aggression toward the Indigenous people in the past, they do not worry about it now, and they will not worry about it in the future. Furthermore, officials show no interest in correcting these impacts. Dirt and heavy metals make the fish in many rivers unfit for consumption, but these communities lack an alternative. The government provides no signage to warn people about the river pollution. Even visitors from other regions unknowingly bathe in these waters, despite their murkiness and the latent threat they carry of harboring unpredictable diseases. The right to live free from pollution becomes more distant every day.

Part 5. The Faithful Remembrances We Must Not Forget to Counteract Future Consequences.

We must close our eyes. I wish we all possessed the gift to remember past stories as if the events happened in that exact moment. Remembering our lived experiences gives us the ability to differentiate between the past and the present, allowing us to correct errors that will surely carry consequences in the future. Ignorance of history increases the apathy of institutions and society, because people ignore the true present situation. This ignorance makes it impossible to define the true goals we must set for the future, or to restore the environments we have damaged and destroyed in each place.

Admittedly, with this background, I nurtured the idea of reversing various actions throughout my life—the very actions that people executed globally, causing severe environmental and social impacts. Consequently, in 1980, I began the great challenge, even though completing it would require the rest of my existence. Restless in the face of growing ecological disasters, I decided it was time to begin researching. I wanted to grasp the possibility of achieving the survival of threatened flora and to confirm the possibility of bringing fauna back and recovering species that had already disappeared from the area. I always noticed that pollution already tainted some of the majestic rivers where we used to bathe and enjoy ourselves. Furthermore, their water levels had dropped extremely low compared to their levels just a few years earlier. As a result, I began to feel like a prisoner in a city that grew at a dizzying pace.

The Goal of Ecosystem Restoration

My primary goal involved discovering a way to return to the world the exact nature humanity stole from it, and to thoroughly research ecosystem restoration. I hoped my results would serve as the best indicators in the future. I actively sought and applied real concepts regarding the rights of Nature, emphasizing freedom through a profound respect for every being that inhabits it. To achieve this, I created environmentally friendly activities to sustain human life without harming nature. I wanted this green island I planned to rebuild to serve as future inspiration for those inclined to participate in ecological restoration for conservation purposes. Ultimately, I aimed to establish true natural habitats to ensure the survival of flora and fauna.

The Urgent Need for Active Conservation

Conservation remains necessary and imperative. Buying forests for conservation provides a solid methodology to prevent their destruction. However, we know that these forests located on private properties, or the reserves governments protect in each country, remain tiny compared to the destroyed spaces; they simply cannot counteract global ecological impacts. For this reason, we face an urgent, necessary, and imperative duty to actively restore the damaged environments using the exact vegetation we stole from them.

Part 6. Difficulties that Delayed Certain Activities in Ecosystem Reconstruction.

We must not forget historical events. By remembering them, we can develop proposals to correct past mistakes. In many cases, we can apply the most appropriate logic to favor nature and life in general.

I must describe some of the conflicts I faced during my life because they reflect twenty years of harassment directed at me and towards nature. Individuals and institutions led these attacks. They generally hold the power to destroy both people and nature.

In 1996, the Ecuadorian government created a new state institution. This agency emphasized controlling and sanctioning flora and fauna trafficking. It did not prioritize nature's rights, of course, because society had not yet granted nature any rights. Thus, in 1998, the regulatory institution began persecuting me. Initially, officials repeatedly visited my property. Acting with great arrogance and superiority, they insulted me for planting trees and threatened to cut them down. They imposed a retroactive fine spanning eighteen years simply because I planted trees. I rebelled and refused to pay.

A few years later, they harassed me by threatening to rip the epiphytic plants off the trees. Even though the Ecuadorian Constitution enshrined the rights of nature in 2008, authorities ignored the law and continued their threats for twenty years. Therefore, officials prohibited me from taking any action to continue my research on ecosystem reconstruction. They stopped me from gathering seeds in living forests, logged forests, or anywhere else. They forbade me from working in the communities. They even confiscated the bouquets and anthurium flowers that provided my livelihood. They claimed that if they caught me performing any of these activities, they would subject me to prison and fines.

The harshest blow came from their continuous threats to cut down the entire forest and confiscate the plants making up this still-developing natural space. They argued that moving the seeds or plants I used to build the forest constituted a commercial activity.

However, I actively avoided encountering them during my daily work. I pursued my ideal despite the continuous visits and threats from this Ecuadorian state institution. I will never forget the exact words they used to threaten me:

> "You cannot collect seeds or plants from logged forests, even if they sit on private property. Let them rot. You must not care if any species goes extinct. If you take a single seed or plant, you will face imprisonment and fines."

This persecution lasted twenty years. Finally, in 2017, a director of this institution in the Pastaza Province clearly understood the importance of rebuilding nature. At that moment, the agency granted me the respective patent (I provide details in Chapter II).

Because of this situation, during those twenty years of persecution, I could not consistently advance certain activities until I overcame this obstacle. Due to this and many other institutional failures in recent years, the threat of extinction for important flora and fauna species has increased considerably. Furthermore, oil and mineral extraction policies leave soils and rivers contaminated. These practices inflict irreversible impacts on the habitats of various Indigenous communities.

Applying the law of nature's rights also means respecting the life of every being within it, as every creature plays a vital part in the food chain.

Today, I can state that rebuilding this jungle represents my most valuable life experience, because anyone can physically touch and see the REBUILT ECOSYSTEM. Thus, the following sections describe the tasks I undertook to obtain diagnoses. The physical work and results will form part of the methodology if necessary. Alternatively, future restorers can remove certain actions if those actions do not apply to other restoration projects. Project leaders must adjust all these steps according to the era, the region, or the specific characteristics of the designated restoration site.

CHAPTER II: METHODOLOGY FOR RECONSTRUCTING ECOSYSTEMS IN DEGRADED AREAS

Part 1. Historical Abundance Increasingly Distant to Find

We achieve true affection and respect for nature only when we understand the importance of each of its components. Attentive observation of the smallest details in every being's life reveals the essential functions that sustain the equilibrium of ecosystems. Only by recognizing their value as living beings do we understand that this group of inhabitants—sometimes feared, ignored, or trampled under the false idea of their irrelevance—forms an integral part of our own lives.

So much so that, as time passes, we can no longer easily contemplate the biological and forest wealth that once reigned abundantly in many places. We feel admiration when we find rare, beautiful, and increasingly scarce forms of life. We often fail to notice that they represent vestiges of the natural paradise that once extended fully across the land. For those of us who remember those stories, the absence of these species generates melancholy, sadness, and concern, as we understand that biodiversity faces extinction at an accelerated rate. Faced with this reality, everyone holds the responsibility to restore at least our closest environments. Through this action, we would prolong the existence of countless beings that constitute the very strength of life in general.

The Inception of the Research

The year 1980 marks the beginning of this research. At that time, I found no references or sources outlining a specific methodology for ecosystem reconstruction. Therefore, the path I took relied heavily on creativity, supported by a deep analysis of biological succession processes and the ancient relationships between living beings. Together, these beings formed the extraordinary equilibrium of biodiversity within ancestral ecosystems. Furthermore, memories from my childhood nourished my initial motivation.

I chose three main components as a foundation of hope—a hope that ceased to be a simple illusion and ultimately became a mission fulfilled within this Amazonian research. The chosen elements never stood apart; therefore, they naturally integrated as part of the dynamic process of ecosystem reconstruction and revitalization.

The challenge seemed impossible. The selected area displayed total destruction and loss of flora and fauna: the entire space consisted of pasture, serving as a glaring footprint of livestock expansion. But my heart beat with strength and hope. I simply exclaimed my wish that this forest-building process would serve society. I hoped that when the trees eventually shared their shade, they would possess the extraordinary strength and health to offer people the opportunity to look, listen, appreciate, and breathe deeply those airs of freedom.

And then, perhaps, I could successfully bring back the animals and birds that long ago inhabited this place but had disappeared due to inhuman actions. Furthermore, I wanted the chance to support society in increasing its awareness, love, and respect for nature, offering people the opportunity to see, appreciate, listen to, and breathe freedom.

The Core Hypothesis and Long-Term Vision

To achieve success in this endeavor, I had to firmly apply the concept of freedom and respect for the life of every being that would progressively return. The challenge consisted of proving that the disappeared fauna still had a chance to return, even though the initial scenario seemed to deny it, because only pastures existed in the chosen restoration site and its surroundings. Of course, we lived in the era of cattle ranching; grass held immense value, and people gave forests no importance. Creating those pastures had required immense labor and a high economic cost to eliminate the jungle.

Instinctively, I desired and believed that future results must be positive. However, I always remained aware that the long-awaited return of valuable indicators would take many years. These indicators would ultimately validate the hypothesis and, surely, allow me to propose a recommendable and replicable methodology to society.

Hypothesis: Planting the lost flora will bring back the lost fauna.

I clearly understood that this represented a long-term challenge. I fully grasped the magnitude of the task, because obtaining result indicators would take a lifetime. Even so, excitement flooded my mind as I imagined that this methodology would allow us, for the first time, to maintain living germplasm banks containing endangered flora species. I envisioned that we would manage to establish original habitats for flora and fauna, and that we would discover strategies to achieve the survival and equilibrium of reconstructed ecosystems.

Additionally, the project would demonstrate that restorers could achieve the strength of restored ecosystems in less time than natural succession requires. With the progressive return of pollinators, the vegetation would gradually produce its own seeds. Birds, animals, and the wind would then disperse these seeds. In this way, I planned the manual reproduction of various species and scheduled vegetation donations to local communities. I hoped to transmit the commitment to conserve the most valuable natural specimens until we reached the equilibrium that nature requires. This process would achieve the resilience necessary for the ecosystem to survive without human manipulation, thus consolidating a true ecological and social reconstruction.

Motivation: Perseverance and positive stubbornness provide the strength for success.

Part 2. Elements Designated for Ecosystem Reconstruction

In 1980, people still found it unusual to discuss elements intended to regenerate ecosystems. Furthermore, society did not recognize the importance of conserving native species. Instead, institutions prioritized introducing exotic species like pine and eucalyptus in the Andean region, teak in other regions, and pastures in the Amazon. Decision-makers considered fast growth for timber harvesting as the only criterion. Under this logic, the original vegetation disappeared at an accelerated rate, while people planted native ecosystem species in minimal and insignificant quantities.

This research proposes that restoration must be comprehensive, because the proposal pursues conservation rather than timber harvesting. To achieve this, I have defined three essential elements. If we omit any of them, the results will lack the true strength of an Amazonian ecosystem. At first glance, certain sectors of society viewed this vision as a utopia. However, the Cultural Element remains indispensable. It constitutes the science of ancestral knowledge regarding the value of forest and biological resources. For generations, this knowledge allowed Indigenous peoples to use these resources without causing destruction or deterioration.

I consider three elements to restore Amazonian ecosystems: the Forest Element, the Biological Element, and the Cultural Element. I analyzed each of them before applying them to the chosen area.

At that time, people predominantly practiced locking wild animals in pens and cages according to their convenience, justifying this action by claiming it protected the animals. At the same time, hunters and colonizers widely adopted the custom of killing any living being that moved. In other words, society lived in an era that excessively exploited majestic nature's resources.

The experiences and tasks I describe in the physical works constitute an essential part of the methodology I applied to reconstruct ecosystems. My efforts and the results I obtained will not remain solely in written records. Instead, they materialize in a living, real, and palpable reconstructed ecosystem. This environment allows visitors to form their own opinions regarding the authenticity of this life story.

Site Designation

In 1980, I purchased a 6.75-hectare plot of land covered in livestock pasture, located 3 kilometers from the city of Puyo. I intended to reconstruct a living ecosystem that could become a model for restoring degraded areas in the following years. The soil already showed clear signs of erosion. The presence of acidic, nutrient-poor clays was evident, a common characteristic of most lands occupied since the Colonization era. I considered the space ideal for two reasons: its apparent distance from Puyo and the fact that I needed to conduct the research precisely in locations where forest and biological damage had been most severe.

I remember that only three families lived along the road leading to the chosen site. A tranquil and silent atmosphere reigned, broken only by a stone and sand road that bordered one side of the property. A vehicle would sporadically pass by heading toward the Pindo Grande River, where people used to bathe or fish. These activities remained worthwhile at that time, as the waters flowed abundantly and remained clean.

I never imagined that Puyo's urban expansion would reach me so quickly. Back then, I viewed the site as the perfect space to raise a small jungle. I saw it as a place of peace and tranquility, destined to shelter, in the future, every being that found refuge within that enchanting and still-imaginary vegetation.

Diagnosis of Existing and Disappeared Flora in the Surrounding Areas

Conducting research before beginning physical restoration work constitutes a fundamental step. These inquiries allow restorers to establish an initial and progressive diagnosis, which serves to measure the achieved results. This diagnosis provides the baseline for comparative analysis from the beginning and throughout the ecosystem's development process.

Recovering the historical memory of the site's flora remains essential, as this allows us to identify the moment the ecosystem begins to reach optimal health. The restoration of the food chain complements this process. This restoration reflects itself in the progressive return of fauna, whose individuals act as natural indicators of recovery. Thus, the survival of the planted species will mark the moments when the forest achieves a certain level of sustainability. This makes it easier to evaluate the results, determining whether they are positive enough to continue or negative enough to require correction.

The next step involved surveying the surrounding lands. By observing the vegetation remnants, I identified the species that ancestrally populated the sector slated for restoration. During this process, I found it indispensable to converse with elderly people, former owners, or residents of nearby areas. They provided valuable information regarding the vegetation that existed in the past. Through this dialogue, I discovered that humanity had completely eliminated a large part of that essential nature.

This exercise allowed me to diagnose the species that, in ancestral times, constituted the ancient forests of the restoration site and formed part of the Ecuadorian Amazon's green component. Furthermore, it provided a guide to identify the basic vegetation and set the challenge of obtaining seedlings or seeds of these species, even if I had to search for them in distant locations.

In this way, I could fulfill a primary principle of nature's rights: restoring the area with the exact vegetation that human action had exterminated. Although humanity completely eliminated these species from the chosen lands, they had once thrived there.

Restorers can compare the future results obtained from the restoration process with the initial diagnosis built from observation or inquiry. These results may vary according to the location, the time elapsed, or the level of destruction in the intervened space. However, even if the collected information is not exact, it remains essential. It provides highly valuable approximate background data that will guide the restoration. In this regard, information provided by elderly Indigenous people, by those who inherited knowledge from their ancestors, or by early colonizers who closely knew the original nature of these lands holds special importance.

Reconstructing an ecosystem provides the opportunity to plant species from different locations, although restorers must fundamentally use species from the same region. Furthermore, one can include species not considered during the initial inquiry. Far from negatively affecting the process, this inclusion enriches the possibilities of favoring the return of fauna, particularly species that migrate to other territories due to climate change in search of new conditions to adapt and survive.

Diagnosis of Terrain Conditions for Restoration

Observing the extensive pasture destined for cattle felt entirely discouraging. Eliminating the pasture seemed like an impossible mission. However, perseverance, firmness in my objectives, and the conviction to achieve a goal boosted my spirits. My mind filled with ideas and images of what might eventually reveal itself in that still-uncertain place. At first glance, eliminating the grass across 6.75 hectares seemed impossible. I knew the task required dedicating part of my life, but curiosity about the unknown fueled my motivation. Deep down, I repeated a phrase that accompanied me throughout the process: *"What is never started, is never finished."*

I found it essential to observe the soil conditions and meticulously understand the degree of topsoil deterioration in the chosen space to find strategies to improve its quality. In this case, I found the soil deeply eroded due to the terrain's slope, the constant trampling of cattle, and the heavy rainfall characteristic of the zone.

When the terrain presents unevenness and slopes of varying magnitudes, one can calculate the degree of historical erosion and anticipate what might happen during the restoration process. This observation allows restorers to foresee scenarios and design adequate alternatives to prevent further deterioration of the topsoil that will progressively build up throughout the process.

Diagnosis of Flora Contents at the Restoration Site.

This activity involves identifying the vegetation already present at the chosen restoration site. Restorers will likely find a tree, palm, shrub, or even grass that, while appearing insignificant, may play a crucial role in the process. Small plants, for example, can attract butterflies, hummingbirds, or other insects; therefore, restorers must not eliminate them. We must fundamentally understand that every element of nature holds significance within certain abundance ranges; whenever possible, restorers should exclude nothing. Each organism can prove vital for the first insects, amphibians, or other small animals that already inhabit the area or will arrive over time. For this reason, I proposed conserving everything possible during this stage, recognizing that certain species might become difficult to find later. By keeping them within a restoring ecosystem, we facilitate their observation, conservation, and propagation.

During this observation, I remember finding, hidden in a small, hard-to-reach grassy hollow, one Pambil palm (*Socratea exorrhiza*). In another sector, I found four Chonta Pambil palms (*Iriartea deltoidea*), and three Kili palms (*Wettinia maynensis*). These palms exceeded fifty years of age and proved fundamental for collecting and reproducing seeds. Notably, local experience taught me that these palms require over two hundred years to produce incorruptible wood suitable for construction. Furthermore, they only thrive in fertile soils; under adverse conditions, they remain small with thin stems and often age prematurely, dying from a lack of nutrients.

In another sector, I also found two Tachuelo trees (*Zanthoxylum sp.*) and two Tamburo trees (*Vochysia sp.*). Although they do not produce fine wood, I considered their conservation important, intuitively knowing they might face extinction in the future. Likewise, I found one Guayacán specimen (*Handroanthus chrysanthus*), a species already threatened with extinction that I understand requires at least three hundred years to produce high-quality wood.

These few species proved decisive, contributing their seeds for propagation and ensuring their survival throughout the restoration process. Aside from these isolated cases, cattle pasture and a few scattered Pigüe trees (*Piptocoma discolor*), which I had to cut down, covered the remaining 6.75 hectares.

One can sporadically find the tree known as Pigüe in primary forests, although it commonly appears in eroded areas following landslides or flooding. For this reason, Pigüe naturally emerges immediately after the logging of primary forests, reproducing easily in clay soils. In fact, it acts as an indicator species. However, once a restorer successfully restores the topsoil, these trees age rapidly and die.

At that time, due to ignorance, finding them excited me. I thought they would develop and prove useful in fertile soil. Indeed, they grew for twenty to thirty years. But upon reaching maturity, their roots rotted, and they died. Consequently, upon falling, many crushed small trees of valuable and endangered species. Furthermore, they provide poor and limited biomass and nutrients to the soil. Therefore, when starting a restoration, restorers find it more convenient to eliminate this species, as it can damage the planted vegetation in the future. Notably, Pigüe barely exists in primary forests. Today, loggers use its wood to manufacture fruit packing boxes.

Monitoring of Wetlands at the Restoration Site

In the Amazon, restorers almost always find small or large wetlands or springs, some highly deteriorated, especially in pasturelands. This component holds utmost importance for the return of amphibians or fish. Furthermore, the quantity and quality of progressively restored water constitutes another essential indicator for evaluating results, potentially becoming a measure of the forest's health.

Wetlands also contribute to the return of more sensitive aquatic beings, depending on the water quality and wetland size. Additionally, restorers must reforest these areas with aquatic plants, palms, or trees, which improves the environmental quality in these settings. The surrounding vegetation acts as a filter and progressively feeds the wetlands. However, this depends on the forest's age, the size of the restored land, and the biomass recovered within it.

When abundant vegetation surrounds wetlands, the water usually remains crystalline and of good quality. Conversely, when swamps or small streams remain exposed to the sun, the water acquires a yellowish color and evaporates quickly, causing the disappearance of habitats for countless species.

A swamp existed on the chosen land, becoming an important element for monitoring the sequential recovery of water quantity and quality, and serving as a valuable indicator of returning life. However, after 30 years, Puyo's population growth caused the clearing of scrublands and secondary forests around the area. Consequently, the 6.75 hectares of forest cannot maintain the high water levels recovered during the restoration process. Currently, I observe a worrying decrease in the flow of water from natural springs.

Reconstructing Ecosystems in Pastures or Eroded Soils.

The main purpose of this research consists of restoring damaged, vegetation-less areas by incorporating all plant species, especially those threatened or in danger of extinction. However, creating an ecosystem in a pasture represents a laborious task, although it certainly offers better long-term results. This situation provides the opportunity to plant the most suitable vegetation for each site, holding the initial advantage of zero light competition.

Likewise, adequate soil restoration facilitates positive vegetation development. However, restorers must replace dead seedlings immediately while sufficient light still exists for their development. Once the trees mature, introducing new small trees becomes almost impossible. Light competition limits their survival, causing them to die or, depending on the species, remain dormant for a long time until a falling tree opens a light gap. Notably, few species can adapt to lower light conditions.

During restoration processes, restorers must strictly avoid using chemical products. Chemicals delay the recovery of microorganisms. Conversely, applying organic fertilizers increases beneficial microorganisms in the soil. The diversity of plant species adds to this process; their biomass substantially increases elements that aid the decomposition of organic material and the reproduction of decomposing fungi and fungi in general.

When managing a very extensive restoration area, restorers find it essential to create different ecosystems at appropriate intervals. For example, they can establish nuclei of three to five hectares, separated by 100 to 150 meters (this distance can vary). In this way, when pollinators and fruit/seed dispersers arrive, the vegetation will spread into the un-restored intermediate spaces, eventually forming a single ecosystem across the designated site. Although connecting these spaces takes more time, restorers find this method convenient for extensive pastures. Alternatively, depending on available economic resources and time, restorers can opt to advance toward total assisted restoration.

Restorers must include the greatest possible diversity of timber, medicinal, cultural, artisanal, and food plants. This creates forests with multiple resources to sustain diverse fauna. Every plant species provides food for some living being. However, I recommend that approximately 20% of the vegetation provide short-term food for birds, animals, and insects to consume.

Restorers should plant timber trees ten meters apart, avoiding monoculture and dispersing the species. They can insert smaller species, like fruit trees, into the reforested forest, always considering each species' characteristics. Due to their leaf size, palms like Morete (*Mauritia flexuosa*), Ungurahua (*Oenocarpus bataua*), and Chonta Pambil (*Iriartea deltoidea*) require wider spacing if restorers intersperse them with trees; when their large leaves fall, they crush small vegetation.

The most convenient method involves planting palms three meters apart in groups across different locations, allowing trees to grow without interference. Later, birds, animals, and bats will disperse the palm seeds throughout the site, allowing them to grow in chosen locations. The ecosystem will also host an extensive variety of smaller plants; restorers must insert them into the forest, choosing the most suitable location for each.

Regarding orchids, restorers can tie them to trees or plant them in the ground, depending on the species. Restorers necessarily must familiarize themselves with each species to understand its requirements regarding light, humidity, and ventilation. Once the forest develops, natural habitats will form where the wind deposits seeds in favorable spots, growing new plants and ensuring the species' survival.

When the forest reaches a certain development stage, birds, animals, and the wind transport seeds to various spaces, even bringing new specimens to the site. In this process, frugivorous bats play a crucial role in seed dispersal, especially for various palm species.

Advantages of Creating Ecosystems.

I have conversed with many people, and I notice a generalized belief regarding environmental reserves or primary forests. People usually think that endangered trees abound in these areas, or that they can obtain seeds or seedlings of fine woods in whatever quantities they desire. The reality, however, differs entirely: trees producing coveted wood remain scarce. For that reason, when loggers cut them down, they disappear without leaving replacements. Local experience taught me that even though many of these specimens reach five hundred or a thousand years of age, restorers often cannot find even a single seedling around them.

This happens because birds, animals, and insects eat the seeds. Additionally, the lack of light, constant humidity, and abundant biomass produced in a primary forest smother the seeds, preventing germination. A similar phenomenon occurs in a restored forest; after approximately forty years, the recovered jungle begins to acquire primary forest characteristics, and the diverse returning fauna eats the seeds.

Furthermore, a restored ecosystem offers us the possibility of concentrating more species in a specific space. This facilitates planning, monitoring, and seed collection, allowing restorers to establish living germplasm banks containing the species selected for planting. Thus, restorers progressively secure seeds to propagate new plants. Later, they can delegate the task of conserving these species to communities or farmers. By dispersing plants across different locations in this way, we guarantee that we truly contribute to their existence.

Creating ecosystems also favors biological recovery in less time. Moreover, this recovery can prove more diverse and abundant, depending on the planted species. During the forest's growth, one observes different development rates, even within the same species, due to factors like varying light, humidity, and, above all, soil quality. These specificities influence the uniformity of vegetation growth.

Over the years, the ecosystem achieves a remarkable biological recovery: fungi, mosses, lichens, and microorganisms return, accompanied by numerous species responsible for pollinating flowers and spreading seeds—all crucial for the natural reproduction of countless tree and plant species.

As time passes, nature's own wisdom will make adaptations and foster the natural reproduction of other unplanted minor species. The wind, birds, or animals will transport them. Notable families include Melastomataceae, Rubiaceae, Piperaceae, Costaceae, and Araceae. All remain indispensable for the ecosystem's equilibrium. A magnificent moment will arrive when each species can subsist on its own, achieving a strength similar to that of ancient ecosystems.

Collecting seeds in any forest requires careful planning. Harvesters must clear the ground beneath the tree, track flowering and seed production seasons, and sometimes collect seeds directly from the tree before birds eat them. For other species, harvesters must wait for mature seeds to fall and collect them from the ground. I recommend performing this task early in the morning and late in the afternoon before animals or insects consume them. In this process, harvesters always face inevitable competition with wildlife.

Monitoring Seed Production and Dispersal in the Restored Site

After many years, restorers witness profoundly significant moments: observing the return of life. This moment indicates that the restoration is largely achieving its objective. The forest begins to sustain itself and maintain its own ecological processes. The increasing approach and presence of pollinators confirm the recovery of biological interactions, while previously disappeared fauna species begin to return. Faunal habitats become fundamental, and a sufficient number of the returning individuals manage to survive and establish themselves. This forms new ecological niches that allow specific life forms to develop.

Thanks to this growing diversity of pollinators, the ecosystem's most relevant trees and plants begin to flower and produce seeds again—a key process that guarantees the forest's natural regeneration.

Restorers experience these moments as a reward for constant effort. Along the way, the return of each organism strengthens the spirit, fills it with hope, and transforms it into energy to continue. Restoration guarantees prolonged vitality for nature and its inhabitants, even though the human cycle ends sooner. This contrast reveals a certainty: while nature still enjoys its youth, our own existence extinguishes. However, we retain the satisfaction of having left a better environment than the one we found.

Preventing Climate Change Impacts

In the Amazon region during the 1980s, local concern regarding climate change did not yet exist; the media barely referenced the impacts already under discussion in other countries. However, warnings began to circulate, announcing the possible extinction of multiple plant species and the risk of rivers drying up. I remember that the Andean region already showed phenomena like droughts and frosts, while the Amazon displayed the first indicators of climatic alteration—unequivocal signs that we were heading toward environmental degradation.

Society at that time failed to grasp the magnitude of this problem. The idea that natural resources were inexhaustible prevailed. But the progressive pollution of rivers, the disorderly growth of human settlements, and deforestation revealed that the paradise had begun to deteriorate and that we would soon experience terrible consequences.

Currently, climate change impacts appear increasingly obvious: extreme heat waves, abrupt temperature drops, droughts that compromise the survival of recently transplanted seedlings, torrential rains that cause soil erosion, and strong winds that knock down countless trees. Therefore, preventing climate change risks remains indispensable, especially on steep terrain. Restorers must fundamentally avoid planting trees in inadequate locations—for example, near homes or infrastructure—because falling trees could cause damage and force preventative logging.

I clearly remember the severe drought that occurred between August, September, and October of 2010, considered the most intense in the Amazon's history. Springs dried up, and water became scarce. During the nights, I found myself forced to dig holes in swampy areas to extract a little water to hydrate the small trees I had planted that year. Rationing became inevitable; I distributed limited water to each seedling, although many failed to survive the prolonged absence of rain.

Those moments brought despair. However, when the rains return, society tends to forget droughts; when the heat arrives, people forget the cold waves and intense rains. This tendency to forget events proves dangerous. We must remember the historical abundance of water in the rivers and the exuberant vegetation; only then can we measure the gravity of what we have lost to this day.

Following massive deforestation in the Brazilian Amazon and continuous logging in Ecuador, Colombia, and Peru, climate change effects have intensified, even altering species' phenology. The usual flowering and fruiting seasons of many trees and plants no longer correspond to their traditional cycles. In this region, I have observed that species that once produced seeds annually may now go two, three, or even four years between substantial seed crops. This irregularity forces restorers to collect seeds at the precise moment they become available, whether to conserve, store, or plant them. The impact proves worrying, as it also affects fauna. Food availability periods change, arriving early or late compared to usual cycles, generating disorientation in both resident and migratory species.

Community Expectations at the Start of the Work

The proposal to transform a pasture into a functional forest ecosystem initially generated strong negative expectations within the community. Many considered the idea madness. At that time, people attributed more economic value to grass than to the forest. The Pastaza Province did not yet project tourism as a development alternative, and people failed to recognize any immediate utility in planted native trees.

I heard discouraging comments questioning the need to plant trees in a territory where, according to general perception, sufficient vegetation still existed. In contrast, carpenters and builders used high-value fine woods, convinced those resources would never run out. They considered the notion that valuable forest species could face extinction completely unthinkable.

At that time, environmental awareness remained scarce, and respect for planted trees practically non-existent. Several times I witnessed neighbors cutting down and taking away the small trees I had planted. On other occasions, they intentionally damaged them to obstruct the restoration work.

Even so, criticism and negative actions failed to deter my ideas. I persisted in my goal to reconstruct a small model jungle for restoration, convinced that the results would eventually become visible and valued. My goal involved forming an ecosystem capable of sheltering threatened species, although I sometimes doubted the possibility of achieving it. However, I understood that persistence and resilience would become indispensable virtues for achieving the purpose: creating an ideal space for the return of life.

Little by little, I discovered I was destined not to decline. Every step brought me closer to proving that ecological restoration remains possible, and that even fauna disappeared from certain places can return if we create the right conditions.

Preventing Impacts Caused by Residents in Surrounding Sectors

To minimize future negative impacts caused by human activities in the lands adjoining the restoration area, restorers must essentially understand the neighbors' socio-environmental dynamics and the nature of their productive practices beforehand. For example, if streams or springs originate on their properties and cross the restored zone, establishing community agreements to guarantee their protection from pollution proves indispensable.

Population growth represents another risk factor, bringing pressures like excessive noise, hunting practices, indiscriminate logging, establishing unmitigated poultry or livestock farms, and, worst of all, industrial agricultural activities that inevitably pollute the air and water. These processes alter insects' eating habits, disconnect them from their food chains, and turn them into pests.

The most effective strategy involves engaging neighbors in the restoration cause, whether through environmental education programs, training, or by providing seedlings for them to plant on their properties. Identifying external threats in advance allows restorers to plan more effective prevention measures and choose the most suitable restoration sites with greater judgment.

Planting trees requires few precautions beyond avoiding fall risks on homes or infrastructure. However, reconstructing ideal ecosystems demands a broader and more responsible analysis and perspective. Restorers must consider the lives of multiple faunal species that will arrive at the site seeking food and shelter. If restorers fail to take these precautions seriously, they risk causing setbacks; the most sensitive species will disappear again.

Restoration in Secondary Forests.

When a restorer chooses to create an ecosystem within a secondary forest, the initial diagnoses mentioned above remain equally indispensable. The difference lies in the fact that these spaces already possess significant vegetation cover. Therefore, restorers must start with a detailed floristic inventory to evaluate the quantity and quality of existing vegetation. Likewise, they must identify which species the previous occupants completely eliminated or which ones remain absent, to determine which species to incorporate through directed reforestation.

Despite their appearance of abundance, secondary forests can exhibit reduced structural and floristic diversity, particularly after severe or repeated disturbance, and natural recovery of some old-growth characteristics can take many decades or longer. Generally, pioneer or secondary species dominate these forests. The abundance of individuals from the same species generates competition, making it necessary to perform selective thinning. Restorers should leave specimens distributed in different locations to avoid negative impacts on the regeneration and development of new species. Opening clearings generates light gaps that favor the development of the trees chosen for integration alongside the existing ones.

Due to their height or leafiness, naturally regenerated vegetation in secondary forests competes intensely with planted seedlings, hindering new species' survival and limiting the prompt return of fauna. However, by incorporating key species into the restoration, restorers regenerate specific trophic resources (flowers, fruits, nectar, tender leaves). This attracts birds, mammals, and insects, which in turn disperse seeds, creating optimal conditions for consolidating a balanced ecosystem.

Thinning secondary species does not mean eliminating them entirely. All vegetation fulfills a function in regeneration: it produces biomass, prevents erosion, and generates shade. Reducing individual plants simply aims to provide space and favor the development of species possessing higher ecological value. Restorers must complement this work with prolonged monitoring for the development of most seedlings—at least five to ten years—until they reach an adequate size to survive independently. During that time, restorers must ensure the seedlings' access to light and minimize competition with surrounding species.

Dead or removed trees provide essential nutrients to soils and new plant generations. One recommended technique involves girdling or debarking the stem, which causes the tree's progressive death and allows its gradual decomposition, reducing the impact of its fall.

At the same time, whenever possible, I recommend applying localized organic fertilizers to each seedling, especially during its initial development stages. Restorers must decide to fell a tree only after understanding the surrounding vegetation and the possible damage its fall could cause to planted species. Whenever possible, they should direct the fall to avoid direct impacts on growing individuals.

Many trees that naturally colonize secondary forests, such as Guarumo (*Cecropia peltata*, see Figure 12) and Pigüe (*Piptocoma discolor*, see Figure 13), complete their life cycle around 20 or 30 years. When they fall naturally, they can destroy seedlings or young individuals of highly valuable species. Therefore, understanding these species' characteristics proves essential to anticipate future problems and guarantee the restoration process's stability.

Figure 12 - Guarumo (Cecropia peltata)

Figure 13 - Pigüe (Piptocoma discolor)

Reconstructing an ecosystem within secondary forests therefore requires a different set of actions. One could describe it as complementing and replacing trees—that is, replacing the most common and abundant trees with ecologically more important trees that might face threats or extinction.

THE BIOLOGICAL ELEMENT

This element details the biological diagnosis regarding the insect species that existed at the moment I began reconstructing the ecosystem.

Observation of the Existing Fauna on Site

When greater environmental destruction exists in an area, observers likely see a larger quantity of insects turned into pests. This phenomenon also depends on the region, the climate, and the specific elements within the spaces chosen for restoration. Therefore, observation plays a crucial role, as it allows us to make informed decisions to counteract the damage these insects would cause to the first plants.

I do not recommend using insecticides, because we will never manage to eliminate the insects that damage the vegetation. On the contrary, pests progressively acquire resistance. This forces people to use insecticides permanently, killing other, more fragile insects. In the future, these fragile insects would naturally achieve biological equilibrium, thus recovering the food chain.

Success lies in changing the eating habits of the insects we classify as pests. The arrival of new insect species depends entirely on the vegetation we gradually plant. This process provides a perspective to solve the problem, as these insects will once again form part of the food chain. At this point, we can determine that pests do not actually exist in nature, since every organism fulfills a beneficial function in the environment. They become pests because environmental deterioration alters their life systems. Without any food source other than monocultures, we can intuit that they simply seek ways to survive, establishing overpopulations of the same species in most cases.

In 1980, I lacked the facilities to study a specialty that matched my passion for nature. Other regions offered an agronomy specialty, but I found it difficult to access. Therefore, I am not a taxonomy specialist. At that time, knowing the name of each insect or animal held no importance for me; I only wanted to know that they existed and that they stayed alive.

I spent over twenty years photographing all the small animals that progressively arrived. Back then, this record held absolutely no importance. However, when Ecuador established universities offering specialties in Ecotourism, Tourism, Biology, and Environmental Studies, I felt encouraged. The recording and observation of all the arriving creatures also began to gain some importance.

Consequently, university students began to arrive. I remember the first students from the ESCUELA SUPERIOR POLITÉCNICA DE CHIMBORAZO (Adrián Bayas, Santiago Flores, Elena Ochoa, Sofia Secaira, Milka Quingatuña, Gaby Banegas, and many more). At that time, they asked me to let them learn, apply their knowledge, and collaborate in different activities through internships or the preparation of their graduation theses. I viewed this as an opportunity to learn from them. Thus, I learned some insect names. Furthermore, I realized that libraries hold countless names of animals that have already gone extinct, or that people now struggle to find, such as the numerous beetles my brother and I used to play with when we were children.

Figure 14 - Rhinoceros beetle (Dynastes sp.)

Observation of Existing Fauna and Inquiry About Disappeared Fauna in the Surrounding Areas

Acquiring a diagnosis of the historical fauna in the sectors surrounding the restoration area served as the fundamental guide to progressively measure the results and evaluate whether we achieved our proposed goals. If we do not know which birds inhabited or which animals lived in the area, we will have no idea what we want to bring back during the process of reconstructing nature. This knowledge proves vital, even for planning the planting of their preferred foods.

In those early days, observers rarely found birds or insects beyond crickets. Pastures and small plantain crops dominated the nearby lands. Therefore, speaking with the local people proved highly beneficial. The oldest residents remembered the presence of birds and animals they used to hunt to feed themselves, or to obtain skins and feathers for their garments. They also recounted that, in the past, large forests covered the entire region, but their parents or grandparents had cut them down to sell the wood and survive.

Consequently, the restoration's success also depends on the environmental strength of the neighboring lands. Restorers must carefully choose the restoration site and determine, for example, if biological corridors exist, or calculate the distance to forests, reserves, rivers, lagoons, or swamps. This information allows us to project which species we expect to return and estimate the time they might take to do so, considering the life cycles of each organism. Thus, knowing and leveraging external strengths, while simultaneously preventing threats, helps increase the chances of success.

When restorers dedicate a large space to restoration, they achieve more optimal macrofauna recovery results, as the animals find a place to live and reproduce. Conversely, if the territory remains small, even with remarkable recovery, it only becomes a feeding ground without offering conditions for permanent residence. Nevertheless, if conditions remain adequate, microfauna can arrive in abundance. We must remember that insects act as pioneers in forming the food chain.

The Testimony of Don Camilo Santi.

Speaking with Mr. Camilo Santi provided one of my most enriching experiences. He is an Indigenous inhabitant of the San Jacinto community and the ancient ancestral owner of the site I had chosen to restore. Through his lucid memory and soft voice, he shared memories loaded with history and emotion.

With a serene gaze, he told me that large monkeys and caimans in the lower-lying areas once inhabited those territories. Macaws and parrots sang at dusk, wild guans provided them with food, and herds of collared peccaries approached their homes. He also told me, "Further down in the swamp, you will find deep holes; anacondas used to live there."

Don Camilo looked at a very black puppy that accompanied me and asked me to sell it to him. He explained that when he traveled to the jungle, he needed a dark-colored dog: the *sacha runa* (jungle man) cannot see the dog at night, but the dog detects the *sacha runa* and barks, alerting Don Camilo to his presence.

Those words filled me with amazement and nostalgia. However, observing the conditions of the restoration site and the neighboring properties, I thought these animals would find it impossible to return. At that moment, they only survived in nature reserves or very distant places.

Years passed, and to my surprise, several of the animals Don Camilo mentioned began to appear: large caimans, collared peccaries, and an anaconda, among others. But the limited space and external impacts prevented them from establishing themselves as inhabitants of this small jungle and staying for a long time. They arrived, indeed, but they had to leave soon after, exposing themselves to hunters along their way. I remember the case of the anaconda that people tried to kill with dynamite; fortunately, it managed to escape and disappear.

I realized that insects, birds, and animals can perceive appropriate places to live from a distance, but they cannot predict the extent of the territory or its limitations. Even so, their arrival provides a fabulous indicator that we can achieve the return of lost fauna. We only need to transform society's education toward respecting life.

Biological Conclusion: "If you plant the extinct flora in a certain place, then the lost fauna will return."

THE CULTURAL ELEMENT

In the cultural element, I can relate my memories regarding part of the civilization process imposed on Indigenous nationalities.

The Domesticated Monkey: An Allegory of Cultural Dispossession

Real events: A monkey in the jungle forms part of the food chain. He consumes, hunts, reproduces, raises a family, disperses seeds, and regulates others while others regulate him. But he remains free and knows exactly how to live in the jungle.

Humans domesticate the monkey. They put him in a cage or tie him with a rope, feeding him household food: bread, cookies, rice, soda. The monkey learns to shake hands, play tricks, and perform other tasks. He appears happy, but he has become submissive and entirely depends on the food people provide. Now, in the best-case scenario, owners release him; he begs for food, and they must take him for health checkups. Unconsciously, he destroys his owner's house. He breaks the tree in the patio, tears up the plants, and ruins the clothes his owner hangs to dry. In other words, the monkey no longer knows how to interact naturally. He can no longer enter the house because his large size guarantees he will damage everything. The monkey remains a highly intelligent being, but he only knows what humans taught him during his domestication.

Thus, he forgot how to live in the jungle. He depends on humans for food and a place to sleep. Furthermore, humans have destroyed and polluted the nearby jungle. He will surely die if people do not feed him or if he does not discover a survival alternative completely different from his original way of life. This happens because humans erased his knowledge and his freedom; they constantly deceived him.

We can rebuild lost habitats and help recover original ways of life, but this process will take a generation. Therefore, the best path forward lies in freeing their ways of life, reconstructing them, and revitalizing them.

The True Wealth of the Amazonian Peoples

The cultural element represents, perhaps, the most delicate and simultaneously the most consequential part of the restoration process. Unlike trees or animals, here we discuss human lives, memories beating in the elders, and histories still breathing within the communities. Therefore, it deserves infinite respect. Silencing the truth would only deepen existing wounds. Yet, the truth also unsettles an ambitious society that believed itself dominant while remaining ignorant of history, usurping knowledge, and stripping away ancient rights.

The Amazonian Indigenous nationalities did not arrive as guests or outsiders in this land: they were born alongside it and merged with it. No one invited them to inhabit the jungle, because they always formed an integral part of it. For thousands—or perhaps millions—of years, their footsteps marked the true trails of this splendid jungle. They embodied nature itself. They never destroyed their habitats; on the contrary, they learned to listen to the trees, read the river, and interpret the birds' songs, the animals' roars, and the moon's movements.

Many outsiders considered their inability to read or write a synonym for backwardness. But each nationality possessed its own science, transmitting it orally from generation to generation through sacred stories. They practiced minimal agriculture because they did not prioritize it: the jungle naturally offered them everything they needed to live. In every plant and tree, they discovered a medicine; in every palm, a tool; and in every star, a guiding path. Thus, they built a profound worldview—a spiritual language shared with the forest, the water, and the air.

In those communities, people did not view the elder as a burden or merely as someone aging. They viewed him as the wise man, the guide, and the one who sustained his people's lives with great wisdom. He knew the secrets of medicine, the creation stories, the hunting and fishing cycles, and the signs of the sky and the earth. People owed him immense respect because the community's entire history lived within his memory.

For all these reasons, we cannot separate the cultural element from the biological or the forestry elements. When we plant a tree in the Amazon, we do not merely plant wood or shade: we plant memory, we plant stories, and we plant culture. Each tree can provide medicine, food, crafts, or rituals; each plant holds a value that transcends the material world.

The Scars of Colonization and Exploitation

However, colonization arrived, bringing its traps and violence. It introduced the idea of submission. Colonizers offered progress in exchange for the jungle. They promised education, roads, and health; but what actually arrived included spilled oil in their rivers, dead fish, disappeared animals, and the slow agony of their territories. Using small gifts, deception, and sweet words, oil companies tore away entire hectares of forest. A singular true intention hid behind those offers of a better life: plundering the jungle.

Evangelization also made grave mistakes. Although teaching people to pray could have provided a guiding light, missionaries transformed that opportunity into punishment and subjugation. They pushed faith by cutting hair, beating bodies, and forcing people to forget their cultural essence. No one considered that each nationality's true science resided in the great jungle surrounding them. Nature formed a core part of their culture, and severing that link tore away their souls.

Today, Ecuador's Constitution declares a Plurinational and Intercultural State. On paper, it looks like an achievement. But in reality, society continues to marginalize Amazonian communities. The State has not prioritized an education based on their worldview, nor has it enacted policies that respect their territories. On the contrary, external forces have pushed these peoples into a social and economic crisis. Many communities lack access roads, markets, and health centers. Paradoxically, the oil-rich Amazon does not guarantee the well-being of its inhabitants.

The situation remains cruel: society declared the peoples richest in knowledge as "poor." True poverty does not mean a lack of money, but the dispossession of their greatest treasure: ancestral science. Outsiders now commercialize that science globally under different names, while the true owners die in oblivion. And with every elder who passes away, an irretrievable part of that wisdom dies too.

So I ask: Who feels the pain when a community's last wise man dies? Who worries that a unique science disappears with him?

A Vision of Recovery

History could have played out differently. With goodwill and respect, the Amazon would have served as the world's great pantry of unique foods, medicines, and knowledge. Humanity would not have needed to pollute it with oil and mining, or clear-cut it for timber. The Ecuadorian Amazon could have become a territory for cultural and scientific tourism—a space providing a dignified life for its communities.

Indigenous peoples preserved the Ecuadorian Amazon jungle for thousands of years. Yet, since colonization began, outsiders have progressively destroyed it in a remarkably short time. Diverse Indigenous nationalities still reside within it. Seven cultures inhabit the Pastaza province: Zápara, Andoa, Quichua, Shuar, Achuar, Waorani, and Shiwiar. Each possesses its own science and way of life. Several face the imminent risk of disappearance, and the loss of unique, undervalued knowledge causes great concern.

Even today, the State and various public and private institutions actively participate in this destruction. They emphasize extractive programs, causing irreversible damage by polluting the land, water, and air.

We must understand that the Amazon still holds abundant resources to generate sustainable community economies. Ignoring this bounty equals falling into a bottomless abyss.

Reversing history remains difficult, perhaps totally impossible, but we still have time to correct our course. We need firm allies, political will, and respect for nature and culture. If we do not act soon, the damage will become irreversible, not just for the Amazon but for the entire world.

We must remember that the cultural, biological, and forest elements do not exist as separate pieces: they form an invisible whole. Harming one hurts them all.

Today, we still pay a high price for the mistakes made by colonizers and by certain Indigenous leaders seduced by personal promises. But hope remains. The jungle, though wounded, stays alive. The knowledge, though stolen, still beats in the memories of the elders and those interested in rescuing it.

I close my eyes and imagine the intact jungle. I see peoples possessing education and health, but without having commercialized wildlife, without polluting their rivers, and without turning culture into merchandise. I imagine respected communities, proud of their science, sharing their wisdom with the world without suffering deception or dispossession. I imagine the Amazon as a paradise, exactly as it should have been and as it still can be, with dedicated effort. I imagine the Amazon exactly as I knew it when I was a child: A paradise.

Maxim: "Criticism is welcome, as long as an alternative to correct the issue exists."

Within this cultural element, I will cite some examples of plants used for medicine, food, building typical houses, weapons, etc.

Figure 15 - Bitter cane (Costus sp.) medicinal

Figure 16 - María panga (Piper umbellatum) medicinal, food

Figure 17 - Cat's claw (Uncaria tomentosa) medicinal

Figure 18 - Papa tuyu (Dioscorea sp.) food

Figure 19 - Papangu (Cyclanthus bipartitus) medicinal, food

Figure 20 - Rolaquimba (Vasconcellea monoica) food

Figure 21 - Chonta/Chonta pambil (Iriartea deltoidea) food

Classifying an Environment According to its Dominant Content

When we choose a location to obtain plant species (seedlings or seeds) and it presents an optimal forest state—meaning forests without significant alterations or signs of future risk—we must not extract or "rescue" the plants, because they have already naturally chosen that site for their permanence. Seeds represent the exception; we can collect them for planned reproduction programs and reintroduce them into other ideal habitats. For example, if we find orchids in a good habitat, we do not need to uproot them. Rescue only makes sense when a plant has fallen and its survival faces immediate risk. Thus, creating an ecosystem does not justify plundering or sacrificing any form of life. We can differentiate the places we visit to obtain the appropriate flora for reconstructing our designated environment. Based on my perception, I define three classes of environments to establish this differentiation: Natural Environments, Humanized Environments and Destroyed Environments.

Natural Environments

I consider natural environments those ecosystems that humans have never significantly altered or polluted. Furthermore, no one has intentionally inflicted negative impacts on their soil, vegetation, or existing fauna. Their natural dynamics dictate when changes occur. We must note that recovery happens rapidly and positively in spaces suffering natural deterioration, such as damage caused by winds or landslides. This occurs because the forest holds seeds and dispersers that maintain life cycles, simply waiting for the opportunity to grow and dominate a certain space. From another perspective, natural environments simply experience landscape changes according to normal incidents within a healthy ecosystem. In these environments, collecting seedlings remains very limited, but we can feasibly extract seeds or collect epiphytic plants that have fallen to the ground and will surely die.

Humanized Environments

These spaces no longer function as virgin ecosystems. Sometimes, many forests appear natural, but their forest content remains incomplete because loggers completely or partially extracted certain timber species, mainly those considered fine woods. In these places, destruction looms imminently. The lack of some plant species causes countless faunal species to disappear; in other words, human actions directly caused the ecological impact. We can include reforested areas aimed at harvesting wood—which generally consist of linear crops to facilitate harvesting—and secondary forests in this environment category.

Destroyed Environments

Primary or secondary forests do not exist in these spaces. Humans completely eliminated them and subsequently replaced them with a monoculture completely alien to the original vegetation. For example, a food plot does not allow the natural growth of secondary trees because farmers constantly weed it. Pastures provide a clear example: cattle ranching creates a severe impact, and permanent human intervention actively maintains the grass at the expense of the forest.

CHAPTER III: APPLICATION OF THE METHODOLOGY TO REBUILD ECOSYSTEMS.

Creation of the "Las Orquídeas" Botanical Garden

A FOREST CALLED A BOTANICAL GARDEN

Society did not warmly welcome the idea of calling a forest a Botanical Garden at that time. In traditional Botanical Gardens, visitors can appreciate plants from different parts of the world; curators maintain many species in adapted climates they create according to the species' needs. Even so, these gardens hold significant importance for education, tourism, and research.

This forest, which I would call the "Las Orquídeas" Botanical Garden, creates a new role. Specifically, this role involves the conservation, reconstruction, and comprehensive revitalization of all regional components: flora, fauna, and cultures.

Performance of Activities and Application of the Methodology in the Designated Area for Ecosystem Reconstruction (Flora, Fauna, and Culture)

FLORA

Elimination of Pastures

The physical work began at that stage. I was only twenty-three years old. I felt enough energy, willpower, and some anxiety to start and observe the results. Although I worked as a bank employee, I had no desire to spend a large part of my life in an office. I dreamed of rebuilding a forest so that, along with the biological diversity it would attract, it would eventually shelter my family, allowing us to share the success as part of our lives. However, I did not plan to build a forest to promote tourism; that was not the initial purpose because tourism did not exist in this region. I progressively eliminated the pasture while simultaneously building the topsoil and searching for plants to grow in this place.

It took me several years to completely eliminate the cow pasture. I used tools like machetes, hoes, and shovels, and over time, I used a brush cutter. During this stage, I realized the soil surface lacked earth suitable for cultivation; it contained only clay. On repeated occasions, my older brother Efraín, who was twenty-five years old, selflessly helped me. Without his help, I would not have achieved several goals. The work was overly arduous, and we had to endure rain and intense sun because the area lacked vegetation. I remember we used a bitter orange tree as our refuge; we quenched our thirst with its fruits and escaped the intense heat in its shade.

Nevertheless, this labor provoked curiosity, as well as criticism and comments from many people. They expressed that attempting to eliminate pasture to build a forest was foolish. They supposedly had their reasons: in 1980, a property yielded high profitability if the owner cultivated it with pasture. Conversely, properties with forests held less value and lacked commercial appeal, considering the expense an owner would incur to cut down the forest and plant pasture. Regardless, I merely got slightly ahead of my time by eliminating what people valued most at the time—the pasture—to begin planting what they least valued—the trees. People still thought that cattle ranching and timber extraction would generate a profitable economy and provide the much-desired wealth in the Ecuadorian Amazon.

Most inhabitants of this area did not yet realize that we had already lost the original forests in many places. Consequently, secondary forests lacking original vegetation surrounded us, and we ignored that we were aggressively killing that nature, which I consider the very soul of the jungle. Finding important flora species was already becoming a luxury, but people continued to hold a mistaken view of the ecological disaster occurring around us. Regardless, they still thought the region's strength lay in extracting timber and replacing the jungle with cattle pasture.

To finance the high costs of these various activities, I cultivated hybrid anthurium plants and sold their flowers to different florists across the country. This business primarily helped finance the reconstruction of the topsoil. I made a very high investment over twenty years, but this effort enabled the planted vegetation to develop. Otherwise, the most prominent trees would not have grown, because they require ideal soils, much like those found in primary forests.

RESTORATION OF TOPSOIL

Recycling Organic Materials.

People frequently referred to timber exploitation as forestry management, considering it a renewable resource. Therefore, large factories increasingly proliferated in the City of Puyo. These businesses prepared wood for local sale, manufactured fine wooden furniture, and transported products directly to other regions of the country for export. I remember that countless factories sat on the banks of the Puyo, Pindo Chico, Pindo Grande, and Pambay rivers. This location made it easy for them to dispose of logs, sawdust, and all wood waste by throwing them into the waters, which caused intense pollution. I decided that recycling these materials to produce compost was a great option. It certainly was a good decision, because shortly afterward, the Puyo City Council passed an ordinance prohibiting people from continuing to pollute the rivers with these materials.

Consequently, workers at many factories filled hundreds of sacks with sawdust to dispose of it. I then transported tons of sawdust daily to the land I intended to restore. Similarly, I recycled sugarcane bagasse. I mixed these materials with organic chicken or guinea pig waste to breed microorganisms and accelerate the decomposition process, transforming the mixture into humus.

This work was arduous, yet the most appropriate for that era. Although transportation and wages for the people helping me raised the costs, I transported around two hundred sacks of materials daily. During June, July, and August, I hired more assistants, and we transported an average of five hundred sacks of sawdust each day. As a result, after twenty years, I had sectors with 10 cm to 30 cm of topsoil. The effort proved worthwhile, considering that from my observations, I estimated that in natural conditions, the biomass generated by dying trees, falling leaves, and branches would take more than a hundred years to form just one centimeter of humus. For this reason, the ancient forests of the Amazon contain only between thirty and forty centimeters of humus, except for a few specific locations that feature thicker arable soil.

As with all my activities, Agronomic Engineers heavily criticized this work, murmuring that I was introducing numerous diseases to the site through the sawdust or sugarcane bagasse. However, it did not take long for them to start emptying factory waste to improve city lands, parks, and avenues. I am pleased to remember that I was the first person to teach others how to recycle this type of waste for soil improvement.

Currently, not many wood factories exist in the city, nor can people find such coveted woods because they face the threat of extinction. Factories produce limited sawdust, which forces buyers to pay money to acquire this material.

Cultivation of Legumes

Due to massive deforestation, one could easily predict that, in a few years, restoring soils with wood waste would become impossible. The simple reason was that many important tree species—which are non-renewable resources—would go extinct, and the wood factories would cease operations. This concern drove me to research a new method to restore soils. Although the results appear more slowly, proper application makes the method effective. In this way, we achieve encouraging results that allow us to improve soils and reproduce ecosystems on a massive scale in different locations.

Thus, using this background, I conducted various trials with several legumes, achieving great success by introducing the guaba bejuca (Inga edulis). This tree develops rapidly and produces abundant branches and leaves, which supply the soil with nitrogen and many elements favorable for the growth of other vegetation. The guaba tree provides ample biomass, and farmers can manage it effectively through pruning, depending on the crops they intend to grow in the area. Therefore, planters place the guaba trees very close together, recommending a distance of 1 m to 1.50 m. They then continuously prune the trees at a height of 1.50 m, which generates biomass and nutrients in the soil. Planters maintain or eliminate this process according to their needs, applying it to agricultural lands or areas where they decide to create ecosystems; they can plant the guabas and other trees at the same time.

After two years, farmers can harvest biomass to cultivate food crops or family gardens. Thus, if necessary, they can maintain the trees through pruning for many years, or they can completely eliminate all the trees; rotting trunks and roots provide nutrients and soften the soil. If necessary, planters replant the area and repeat the process. One must closely observe the dynamics of the vegetation's development to choose the right action regarding these legumes. Planters generally eliminate the guaba trees, either temporarily or permanently, when the reforested trees reach a considerable height and produce their own biomass, thereby generating new nutrients. Furthermore, once they control erosion, it is advisable to leave many guaba trees at safe distances.

Note: One must consider that guaba trees maintain a rapid growth rate, and planters must closely monitor their development to decide whether to prune or eliminate them. Otherwise, they would cause problems for the planted saplings by blocking the light. Similarly, cutting them down when they are too tall causes damage to the planted vegetation.

ACTIVITIES TO ACQUIRE FLORA

Expeditions in Search of Humanized and Destroyed Environments

I felt worried and somewhat hopeless; I wondered what the first plants I should sow would be. Thus, I decided I must begin primarily by looking for plant species already threatened with extinction. I believed this task was urgent because I knew I would no longer find them after a short time. I also had to analyze where and how to find them, as there were no nurseries back then, nor anyone concerned with reproducing those species.

At that time, acquiring the vegetation required to rebuild the ecosystem proved difficult because forest nurseries did not exist. However, loggers were cutting down primary forests to create pastures or extract timber. For me, those places were ideal for obtaining the plant species I longed to plant. Consequently, I had to travel very far. I remember calling these routes "journeys without a destination" because we would start on a road and travel aimlessly until we found recently logged forests. On several occasions, these trips took us over a hundred and fifty kilometers away.

When we found felled forests, we sought out the landowner to ask for authorization to enter and rescue nature. Although some people readily agreed, others refused, and we even received threats. This labor of searching for plants required immense patience. Nevertheless, obtaining different species in these destroyed forests created high expectations for us. We knew we were performing a true flora rescue, and we felt certain we were saving lives—these plants would soon die from sun exposure, habitat loss, and the fact that farmers would occupy the environment with other types of crops.

We dedicated many years to the plant rescue work. We had to travel to various places in the provinces of Napo, Pastaza, and Morona Santiago, and sporadically to other locations in the Ecuadorian Amazon. On these trips, merely looking at nature and seeing gigantic trees felled on the ground horrified me. This scene allowed me to imagine what those forests looked like before loggers destroyed them, and I thought about the possibility of replicating the exact vegetation association found in the original jungle.

Despite the difficulties, I possessed a fierce enthusiasm to achieve what seemed impossible. Persistence gave me the strength to overcome obstacles and walk toward my proposed goal. Thus, at some point, I had already planted a large collection of threatened or endangered species as a foundational step. However, I did not exclude species that seemingly offered no benefit, knowing irrefutably that they were also crucial for creating the ecosystem.

On absolutely all the plant rescue trips, my older brother Efraín unconditionally accompanied and helped me, just as my father accompanied me on several occasions. My brother played an important role in the early years of work, participating in various essential field tasks to progressively achieve the ecosystem's development.

Figure 22 - Plant rescue

Figure 23 - Plant rescue

Observing Flora Distribution in the Forests

During our trips to acquire plants, observing how nature distributed the vegetation proved just as important. This curiosity helped me make crucial decisions regarding how to plant seeds and saplings, combine the vegetation, and achieve success in the development, adaptation, and diversification of floristic species at the restoration site.

We learned something new on every flora rescue trip. **Each zone served as a school.** Looking at each destroyed forest felt like another wound to nature, but simultaneously gave me a powerful motive to continue. Then, when I had the opportunity to stand inside a living forest, I felt immense inspiration. It became my strength and hope that someday I would achieve my desired success: a forest planted with my own hands, accessible so society could experience a small sample of the Amazon.

Criticisms persisted when I planted the first tree specimens. Critics argued that my work was unnecessary, or that I should at least plant the trees in straight lines. They also insisted I should space them equally and group them strictly by species. I always held a firm idea; nothing intimidated me, nor did I allow such judgments to discourage me. I remembered what the jungle looked like before people invaded it, and I felt certain that nature's disorder was perfect—because in the jungle, each species wisely chooses its own living space.

In the forests, some trees grow almost clumped together while others grow far apart, regardless of whether they belong to different species. Therefore, I knew that trees do not grow in rows in any natural forest. My goal was to build an ecosystem with characteristics similar to primary forests, applying an analysis that would allow me to accelerate forest and biological recovery processes. In this way, I became increasingly convinced that, due to the growing and accelerated ecological deterioration worldwide, my greatest critics would soon realize I was right. This would ring even truer when nature gave me the opportunity to progressively demonstrate these positive results to society.

Planting Rescued Seeds and Plants

Seed Germination: When I found seeds, I created seedbeds directly in the ground; at the exact moment of germination, I transplanted them into plastic bags.

Seedling Adaptation: If I obtained small seedlings, I planted them in bags and placed them in shady areas to protect them from the harsh sun. Once they adapted, I planted them in their permanent locations.

Depending on available space, I highly recommend building a simple greenhouse using 50% shade cloth, where you will keep the seeds or plants in bags or other containers. Later, once they reach the appropriate size, you must transplant them to specific locations for conservation. Bare-root plants you acquire from the wild can easily perish from too much sun, lack of humidity, or other factors you must proactively prevent. Currently, however, you can find some important species in private greenhouses across different regions of the Amazon. You can transplant these plants immediately because their roots do not suffer the severe damage they endure when someone uproots them directly from the forest floor.

You must observe and imagine the physical space a tree's branches will occupy when it matures. I advise planting most species ten meters apart. However, you can plant some non-competing species much closer together. You can plant many palm trees sporadically, clustering up to four plants of the same species a meter and a half apart. In this way, you create very striking natural landscapes by exposing their roots, without causing competition among them.

I strongly recommend planting several palm species—such as Iriartea deltoidea, Wettinia maynensis, Socratea exorrhiza, and Oenocarpus bataua—by creating isolated "islands" of these species in different spaces. This layout provides an opportunity for other trees to develop without the massive palm leaves crushing them. Later, when the palms produce seeds, natural dispersers—primarily bats—will scatter them throughout the area, and they will grow in spaces they naturally choose.

Figure 24 - Seed spreading bat

Replacing Dead Plants or Seeds.

After planting the different species, you will naturally find dead plants or seeds due to various environmental reasons. Depending on their ecological importance, replacing them with new seedlings is essential; you can use the exact same species or choose others based on availability. This approach optimizes all available space. Keep in mind that once the surrounding nature has grown considerably and intense light competition exists, you will no longer be able to replace them.

During the ecosystem's development process, even though you plant saplings at the exact same time, some will grow successfully, others will remain small, and some will inevitably die. This phenomenon should not cause concern. In deteriorated soils, even if you actively restored the topsoil, the surface lacks homogeneous nutrients. Consequently, nature will not develop at a uniform pace either.

The ultimate priority is ensuring that enough individuals of each variety achieve optimal growth to guarantee the future survival of all species. If we observe natural forests, we do not see an overabundance of trees of the exact same species. What truly matters is that enough individuals develop so that, in the future, they provide us with their seeds for other restorations, thereby allowing us to obtain the most diverse living germplasm banks possible.

DIFFERENCE BETWEEN REFORESTATION AND ECOSYSTEM CREATION

Reforestation

Traditional reforestation involves planting trees in rows. Depending on the species, workers plant trees at short distances from one another, or group similar species together, and maintain them through a system of branch pruning. Consequently, planners design these areas to facilitate future timber harvesting. Likewise, in most cases, planters reforest using trees from other regions or countries, always calculating economic profitability in the shortest possible time. They generally choose fast-growing trees over native fine-wood trees. Therefore, after approximately fifteen or twenty years, loggers cut them down for timber, triggering the environmental impact once again.

These reforestations generally represent monocultures that fail to generate adequate biological regeneration, and they provide limited or temporary contributions to climate change mitigation. Consequently, in most cases, landowners do not replant trees in these spaces. Instead, they allocate the land for agriculture or cattle ranching, or at best, they abandon the land so a secondary forest naturally appears after many years.

Ecosystem Reconstruction

Ecosystem reconstruction consists of forming a replica of a primary forest. Of course, scientists would not classify it as primary because it lacks the necessary age for that concept. The purpose is to achieve forest and biological results that closely resemble ancestral forests in less time than nature would naturally take. Furthermore, it primarily involves planting the original flora that ancestrally existed at the restoration site. We must treat all possible forest and non-forest species as important elements, considering even the vegetation we classify as weeds to be essential. Although they might lack an apparent utility, they will surely prove valuable to the life of some organism.

An ancient ecosystem maintains its apparent disorder; however, it represents the most perfect aspect of creation. We must imitate this coexistence and associate the species in such a way that they will not compete with each other in future years. In primary forests, many trees stand far apart from one another, while others grow close together, even among different species. Their giant roots protrude, demonstrating how they support each other, while they also share nutrients and energy. If we decipher this way of life, it surely provides a lesson for humanity, demonstrating that society can live successfully for thousands of years.

Properly associating vegetation is crucial to attract exuberant fauna. This gradually generates the restoration of the trophic or food chain, thereby achieving the return of pollinators and seed dispersers. This dynamic guarantees the survival of each plant species.

During these stages of forest development, many planted or naturally occurring tree and shrub species will complete their twenty to thirty-year lifespans and die. They leave open spaces for the growth of small plants or the germination of seeds, which patiently await adequate light to grow. This dynamic is perfectly normal, because each species achieves natural survival this way. Every moment is important because the vegetation regulates itself. Thus, each plant chooses its living space and associates perfectly, which strengthens the planted vegetation; additionally, it creates optimal conditions for epiphytic plants that I came to understand will survive for thousands of years. This does not mean that the dead trees disappear; they merely had a shorter life stage. Therefore, they successfully reached their natural reproduction stage.

In spaces where we create ecosystems, many species take time to adapt, primarily due to soil quality. But as the forest becomes more vigorous, nature forms its most perfect habitats, which become the strength of the biological element. Introducing new species from the same region causes no negative impact; instead, they become an important food source for birds, animals, and insects that will migrate due to climate change.

When we create an ecosystem, we must scatter the plants. For example, if we have three plants of the same species, we should place them at considerable distances from one another. We must investigate the growth rate of each species, ensuring we never associate fast-developing species with slow-growing ones, to prevent future competition between them. In conclusion, no specific order exists for planting them. By following this method, we will achieve outstanding results in natural biological recovery.

Maxim: "In disorder lies the perfection of nature."

INDICATORS IN ECOSYSTEM RECONSTRUCTION

Recovery of Wetlands

I easily remember the relevant contents of the land I chose to restore. A pasture existed, and a trickle of yellowish, muddy water ran through the center of this land. In this place, I could see the accumulation of organic material produced by the erosion of almost the entire terrain. The continuous trampling of cattle, the slope of the land, frequent rains, and the fragility of the soil caused this erosion. Over the years, a rather deep swamp had formed in various places; naturally, its flow increased only when heavy rains fell.

As the years passed, the planted vegetation grew gradually, and the sun's rays no longer reached the ground entirely. The environment recovered its moisture gradually as the vegetation achieved considerable heights. The different levels or strata characteristic of a tropical forest—such as the forest floor, the understory, the canopy, and the emergent layer—were already taking shape. Furthermore, the soil moisture did not evaporate immediately; on the contrary, the moisture filtered down and improved the quality and quantity of the water. Later, new springs appeared, integrating themselves to support the mission of regenerating and sheltering new life inside this forest.

Currently, these crystal-clear waters form part of significant habitats for several amphibian species, including glass frogs or centrolenids, which belong to the order Anura. Because I had associated these fragile frogs with mature, healthy forest environments, observing them here for the first time thirty years after beginning the restoration work became an especially meaningful indicator for me.

Even though the forest is very small, every event and the arrival of any living being holds a positive value. Therefore, here are indicators of what we could achieve if we unite our efforts and raise awareness worldwide to solve global problems that increasingly threaten the survival of various species, including humanity.

Water remains forever the indispensable element for life. When building an ecosystem, we must focus on conserving the swamps. If the area lacks them, but the space is suitable, we can consider creating wetlands. We must remember that years ago, wetlands were abundant in different areas of the Amazon, and the rivers held more water. Today, the water in many swamps has dried up, and several have disappeared entirely. For this reason, we must remember some history to realize that maintaining wetlands is crucial for the lives of many beings. It is also vital for planting and conserving palms and plants that grow better in those places. This process regenerates critical spaces that encourage the return of amphibians and many other animals and birds that call wetlands their home.

Progressive Seed Production

Various rapidly developing species, as well as others that take many years to grow, make up the ecosystem. As the ecosystem develops, plants will successively produce seeds, which guarantees the survival of each species.

The places where we build ecosystems acquire extreme importance, considering that they transform into living germplasm banks due to the abundance of planted species. Therefore, we must protect them and observe them carefully. These areas facilitate the production of seeds from more species in less space. This concentration does not happen in most native forests, despite their immeasurable importance.

Nature scatters plants or trees over great distances in the wild, and these jungles recede further every day. Consequently, collecting seeds becomes more expensive due to several factors:

- Transportation: Seeds are heavy in many species.
- Time: Collectors must maintain a continuous effort over several days.
- Labor: The work requires paying a specific number of people.

Even so, we must consider that the massive amount of vegetation creates abundant biomass that covers the seeds, suffocating and rotting them. Furthermore, birds, animals, and insects know the production seasons for flowers, fruits, and seeds, and they eagerly eat them. Therefore, we must also note that when birds, animals, or bats scatter seeds, this process guarantees the survival of only a few individuals. In many cases, reproduction fails entirely because the environment lacks the necessary conditions for seeds to germinate successfully.

Healthy Vegetation

The successful growth of trees depends on soil quality and proper care during their various developmental stages. Despite providing this care, we cannot achieve homogeneous tree growth in this sector of the Amazon; we easily see a marked difference in the development of all species. Over the years, several trees will maintain their vigor and produce flowers, fruits, and seeds. I must point out that prodigious zones still exist in different parts of the Ecuadorian Amazon, where excellent soils allow trees to grow homogeneously.

We also measure the forest's good health by the quality of the fauna that arrive to feed, as this depends on the food production the forest provides. Once the forest becomes voluminous, it changes the eating habits of the insects; they stop damaging the vegetation as they did initially and return to fulfilling their original ecological functions.

Soil Erosion Prevention

The biomass that the vegetation progressively generates plays a crucial role in retaining moisture. Consequently, the water filters down and stays in the soil longer. The vegetation prevents evaporation, while the moisture accelerates the decomposition of wood and leaves. This process breeds beneficial microorganisms and forms humus. The soil distributes nutrients evenly, the water quality improves, and water levels rise in springs, estuaries, and rivers. Therefore, the ecosystem creates optimal conditions for vegetation development.

In this way, we understand a simple rule: trees maintain good soil, and good soil maintains good trees. When loggers cut down trees, the stumps rot down to their roots, leaving the soil unprotected and causing it to erode easily.

Fauna Recovery

Although I will discuss this topic later in the Biological Element section, fauna recovery serves as the basic guide to determine if we truly achieved the very goal of creating the ecosystem. It becomes the ultimate indicator of the restoration's success.

In tropical forests, we must meticulously observe the entire dynamic of returning wildlife. We can obtain information about changes in eating habits. Through this observation, we will witness the formation of the trophic or food chain. This dynamic guarantees the ecosystem's survival and sustainability without future human intervention. We must understand that if the vegetation lacks quality health and diversity, the fauna recovery will also remain very limited.

PLANTS THAT MIGHT NEED RELOCATION

Depending on each species' importance and fragility, we must monitor them during their developmental stages. Throughout the ecosystem's formation dynamics, natural adaptations, competitions, and modifications will occur, primarily due to constant variations in lighting. These changes will heavily affect certain species that do not yet have ideal habitats established for their adaptation.

Therefore, we must leave open spaces with fewer trees, where greater light allows us to plant them properly. If necessary in the future, we can reintroduce some plants in different parts of the forest. We also expect birds or the wind to scatter their seeds, allowing them to develop in places each species naturally chooses. Relocating some species remains absolutely essential until their habitats fully form.

Here are a few examples of species that require perpetual observation:

- Orchids
- Bromeliads
- Heliconias
- Anthuriums
- Philodendrons (among other species).

In 1980, while performing plant rescue work, we saved over three hundred orchid species from certain death. I never intended to build an orchidarium, as is characteristic worldwide, because keeping these species in captivity is not the best way to conserve them—even though it has become the standard practice for exhibition or commercialization. Instead, giving them their freedom actively contributes to the balance of ecosystems and life in general.

I never intended to keep the rescued plants in captivity; on the contrary, I tried to improve their quality of life by planting them in natural habitats. Creating ecosystems gave me the opportunity of a lifetime to experiment with the correct methodology. However, at that time, conditions remained unfavorable because the land lacked trees to host the plants, and I had no suitable habitats for each species.

I gathered sticks and trunks from the wood factories to tie the orchids to them and save them from dying. Because this plant family is highly significant and fragile, they suffered greatly. Each species requires special habitats, and some need incredibly unique and complex sites to survive. Seeing them burned by the sun, turned into cricket food, and often falling due to decomposing wood caused me deep concern. Their lives entirely depended on the time the planted saplings would take to grow, as these young trees would progressively form ideal habitats for each orchid species.

As time passed, we saw the small saplings grow and provide shade. I moved the orchids to better places, though not definitive ones, because I merely tied them to the fragile branches of the small trees. They fared much better than before, but the small branches would constantly dry out, forcing me to watch them closely so I could relocate them again. Despite their fragility, these plants proved very brave and survived. Thus, when the trunks of the trees and palms finally grew, I could relocate them to their definitive habitats. This relocation work for some orchid species took me about thirty years.

The vegetation's development formed exceptional habitats—places with varying degrees of lighting, humidity, and ventilation suitable for each species. The humidity created conditions for mosses, lichens, and fungi to develop, indispensable for orchid seeds to germinate. The ecosystem progressively attracted pollinators, which currently face their own risk of extinction.

Forty years have now passed, and seeing small orchids reproducing naturally brings me immense satisfaction. When branches fall from the largest trees, I meticulously observe the epiphytes that have grown high above. I frequently find orchids that have already reproduced, and even new species I had never brought in, because the wind transported their seeds here from other places. Most orchid species now show me that they no longer need my help to survive. However, the severe deterioration of pollinator populations means some orchid species need much more time to regain their strength. They rely on the return of insects that once lived in symbiosis with these plants—insects that now also face the threat of disappearing entirely.

Today, one easily sees orchids in orchidariums: very clean and apparently very healthy plants. However, growers contaminate them with chemicals, and the orchids cannot survive alone without constant human attention. Orchids in the jungle survived on their own for thousands of years. They coexisted with mosses, lichens, and insects; they never needed human help. They actively supported the food chain by sustaining their specific pollinators and other insects that depended on the orchids to live.

Orchids produce beautiful flowers. The human greed to keep them in pots or greenhouses for exhibition or trade still drives people to cut down forests, rip them from their habitats, and subject them to captivity. Consequently, enjoying them in their natural state becomes increasingly difficult.

Many years ago, people cut down forests to create pastures or extract timber. In doing so, they exterminated all the vegetation in those places, including the orchids. Today, logging for timber extraction produces the exact same effect. Undoubtedly, the greed to own or sell these plants has not diminished; on the contrary, it continues to grow. Public institutions openly support this trend, and regulatory entities exhibit a complete lack of concern.

Moreover, since 2019, the free sale of orchids, bromeliads, and many other species has surged in Ecuadorian Amazon markets. An institution in our Pastaza province caused this growing impact. Driven by a policy to attract tourists to the region, officials proposed filling the city with wild orchids—naturally, without maintaining proper planning or establishing an orchid germplasm bank for reproduction. I personally offered to collaborate on this project to prevent environmental damage. Far from accepting my offer, they launched a massive advertising campaign to fill the city with orchids.

Consequently, people from various communities heard this advertising. Now, seeing thousands of orchid plants in the markets is bewildering. Whether out of ignorance or pure necessity, Indigenous people uproot these valuable plants from their own jungles and later sell them in city markets or streets. Buyers take advantage of their humility and pay ridiculously low prices for each plant. Resellers often flip them at much higher prices elsewhere in Puyo and in different cities across the country.

In this way, public institutions continue to create severe environmental and social impacts. They promote false expectations in pursuit of economic returns without conducting prior studies to avoid environmental damage. They completely ignore the fact that the jungle plant sellers make minimal profit, while the resellers enrich themselves. Adding to this problem, the State regulatory entity, which oversees this exact area, fails to intervene and stop this market. This trade constantly increases in the Amazon, leaving us no hope that these actions will end or that humanity will amend this cruelty toward nature.

In primary forests, most orchid species reside in the canopy of giant trees, or in the lower areas, depending on the species. Others grow terrestrially, while some species live exclusively in rivers, rooting themselves directly in rocks. Destroying forests leads directly to the loss of habitats necessary for their lives and natural propagation. In most cases, these impacts remain completely irreversible. Nature formed habitats for orchids and other species over thousands of years. If we do not help, nature will take thousands more to form them again, making it impossible for us to admire natural environments filled with fully free forms of life.

The gradual disappearance of these valuable orchids from their natural places breaks and destroys the trophic chain. The different insects that depend on orchids simply disappear alongside them. If we do not leave the plants in their natural places or rebuild adequate habitats to reintroduce them, we will fail to ensure their survival.

Historically, the Ecuadorian State regulatory entity seized abundant orchid plants from illegal trades. However, these actions did not provide even the slightest solution, because officials later delivered these plants to authorized points of sale. Whether through legal or illegal trade, the orchids remained destined for captivity, causing the exact same environmental impacts.

Reintroducing them to nature actively guarantees their survival. We must keep in mind that in rebuilt ecosystems, periodic monitoring and observation remain indispensable, because relocating them often becomes imperative.

Slowly and progressively, pollinating insects will appear, attracted by the flowers' perfume. In certain orchid species, male pollinating insects collect the flowers' fragrances as part of their reproductive behavior. As they collect these fragrances, they also pollinate the flowers. Specific insects also exist for various orchid species, and these insects face a much higher risk of extinction. Therefore, achieving perfect dynamics in the lives of orchids represents a major accomplishment. The plants will begin to produce seeds; their capsules mature in approximately nine months. Upon opening, they release seeds like spores that the wind scatters across the forest.

Later, nature establishes another vital condition: the formation of mycorrhizae or special fungi in the vegetation. I do not know if these mycorrhizae are specific to each orchid species or if they apply generally to all of them. I maintain this question because, in places with natural orchid reproduction, I have not found diverse species, but rather specific ones. When nature deposits an orchid seed where these mycorrhizae exist, it germinates. Otherwise, millions of seeds may arrive but fail to germinate or survive.

Light and ventilation also play a massive role in developing the small plant in these species. If these conditions are not optimal, the plant dies. Naturally, these species required many complex conditions to survive. Nevertheless, they managed to sustain themselves and live in the jungle for thousands of years.

These emblematic jungle species do not grow abundantly because their lifespan heavily depends on surrounding conditions and their specific location. If other epiphytes crowd the host plant or tree—whether from excessive moss or biomass—or if it loses adequate light or ventilation, the orchid rots or shrinks until it dies. But if the conditions for its reproduction already exist, its subsequent generations will survive.

The market for important hybrid orchids continues to proliferate. These plants look very striking and maintain beautiful colors and designs. Most characteristically, their flowers last a very long time. Because of their beauty, they contribute to tourism and commerce, and florists can use them as cut flowers in arrangements. Therefore, keeping these specific hybrids in captivity does not cause environmental impacts.

Figure 25 - Protective spider

Figure 26 - Protective spider

Bromeliads

These species occupy habitats similar to those of orchids; both terrestrial and epiphytic species exist. Thus, depending on the species, some require greater light to develop and to maintain the vibrant colors of their leaves or flowers. Others require less light. Altering these light conditions causes them to decline or die.

If you plant these species within an area to rebuild an ecosystem, you must actively observe them as the forest develops. Growing shrubs and trees will inevitably obstruct the light, which causes the loss or deterioration of their necessary habitats. When habitats change, you should relocate the plants or temporarily place them in controlled spaces until the forest provides the right conditions to plant these specimens permanently in ideal habitats. Bromeliads will produce seeds; birds will scatter some, while the wind will blow away the wind-borne ones. The plants will then establish their ways of life and adapt according to the specific needs of each species.

Bromeliads act as important and indispensable species in ecosystem reconstruction. We cannot ignore them, because they play a crucial role in forming habitats at various heights and within the forest canopy. The water these plants hold inside allows them to create constant humidity at different levels of the forest.

Birds and animals will bathe in and drink water from the bromeliads; these plants will also provide vital habitats for insects, frogs, and snakes. Thanks to the moisture that bromeliads maintain, ferns, Araceae (such as anthuriums and philodendrons), fungi, and orchids will sprout on the host trees. Furthermore, these plants provide highly favorable locations for birds and animals to build their nests. The importance of bromeliads in an ecosystem is boundless.

However, driven by human greed, people have extracted and continue to extract these species from their natural habitats, specifically targeting the species with brightly colored leaves or flowers. Consequently, vendors display wild bromeliads for commercial purposes in the markets or on the streets of Amazonian cities. This practice contributes to another severe environmental impact. It destroys the food sources of hummingbirds, butterflies, insects, and other animals, while greatly diminishing many benefits that ensure the ecosystem's survival.

Of course, various shops sell beautiful hybrid bromeliads that do not cause environmental impacts. Therefore, we should keep these specific hybrids as ornamentals, whether in greenhouses or under other controlled conditions.

Heliconias

These remarkable plants face severe impacts because they require habitats with abundant light. In the forests, you can find them only in very limited numbers and with sparse foliage due to the lack of sunlight. When you plant them in a space designed for ecosystem restoration, they initially develop optimally. However, as time passes and the growing trees block the light, these plants decline. Several manage to survive because birds eat their seeds and transport them to different locations.

I advise keeping them for a while in controlled habitats where you can manage their light requirements to prevent their extinction; you should also maintain them in places with less dense vegetation. Later, natural dispersers will scatter the seeds, and the plants will survive because they naturally choose the most ideal places for their conservation.

In the Ecuadorian Amazon, people commonly saw many varieties in open spaces like pastures, generally along paths and roads. Growers also cultivated them as ornamental plants for the flower trade, and exporters even shipped them internationally as exotic flowers. Today, seeing some species in their natural state is a luxury. Farmers cut down many plants until they nearly exterminated them, considering them pests in pastures or crop fields.

We must emphasize that heliconias provide indispensable food for hummingbirds, as well as seeds for birds and animals. Some species accumulate water in their flowers, creating natural reservoirs from which birds and animals can drink and essential habitats for insects. These species look striking because their flowers boast vivid colors. Indigenous communities used some of these species for medicine, utilized the seeds of others for crafts, and kept others purely as ornamentals.

Anthuriums and Philodendrons

These species develop as terrestrials or epiphytes, depending entirely on the biomass that forms in the forest over the years—whether on the ground or at different levels across the trees and thick branches. They also depend on the presence of mosses, which retain crucial moisture.

Plants of the Araceae family adapt easily to shady places. Many species grow rapidly. Most of their pollinators are not specific, meaning seed production depends solely on the plant's age. Birds and a diverse array of animals transport the seeds and deposit them in various locations, whether on tree branches or the forest floor, allowing the plants to later climb the vegetation as they grow.

Decomposing trunks provide ideal habitats. The constant humidity favors seed germination. A stabilized ecosystem offers numerous advantages for the natural regeneration of the Araceae family. In the forest, you can observe biomass forming even on the tree bark due to decomposing mosses and bark. This rich layer allows several of these plants to change their life systems from terrestrial to epiphytic. These characteristics stand out noticeably in tropical rainforests.

These species maintain their strength when forest conditions remain optimal. We must actively observe and relocate certain specific species that fail to proliferate naturally and require much more specific habitats.

FAUNA

Tools Used

I remember that I had no funding for this work. I bought a small film camera powered by AA batteries, and a flashlight that ran on two larger batteries. Digital cameras and rechargeable flashlights did not exist at that time. I spent a large part of my financial resources to photograph every individual creature that returned to this restoration site. I carried out this work from the time the land contained only pasture until the forest reached thirty-four years of development. Several critics said it was unnecessary and called me crazy. I still remember how the cost of acquiring film rolls for the camera and batteries for the flashlight kept getting higher. Even so, on many occasions, I treated myself to a small luxury by buying a new camera and a flashlight better suited for this task.

I had to develop a photographic record of the insects, animals, birds, and other wildlife that existed since the land was just pasture, because it provided **concrete evidence of the progressive results** at the site.

At the beginning, crickets dominated the area in sheer numbers. I considered them the initial pests because they fed on the first plants I sowed. The crickets severely affected the orchids, eating their new roots, their shoots, and almost always their flower buds. When I walked through this place, I could easily see crickets jumping from one spot to another.

I soon realized that most insect and animal activity occurred at night. Back then, I lived with my family in the city of Puyo, about five kilometers away. So, I would visit this place around eight or nine at night to observe and learn about the magical ways these insects lived, and to photograph the new insects that kept arriving. Naturally, the progressively planted vegetation actively attracted them.

During this work, I regularly found several insects eating the growing plants; I caught them in a container and released them far out of the plants' reach. Sometimes the situation became uncontrollable, making it seem like the vegetation would never survive. However, I meticulously observed each little animal and realized that, although they caused disasters, they possessed striking appearances, beautiful designs, vivid colors, and symmetrical shapes. Therefore, I never intended to kill them.

Nature served as my greatest school, where I learned many things I had previously failed to understand. Ultimately, acquiring this knowledge only becomes possible when you actively observe and live magnificent moments alongside these creatures.

Formation of the Trophic or Food Chain

There were moments when insects ate important plants, and the situation seemed uncontrollable. Then, it occurred to me to buy grains of corn, peas, beans, and other seeds; I threw them all over the place. As these plants grew, they became the insects' preferred food. Consequently, the small trees and important plants suffered fewer attacks and developed successfully. Later, new species continuously appeared, displaying different and original ways of life. These were also crucial moments for understanding how nature establishes the balance between insects and the environment, and I learned that pests were not actually pests.

Thanks to the development of vegetation and the increase in forest species, various insects gradually fed on other insects, changing their eating habits. Observing these new life forms awakened hope in me; the insects regulated each other, providing clear evidence of the successive formation of the trophic chain. At this point, the plants suffered less damage, confirming that no insect or animal is a pest. In my experience, pests do not exist naturally; humans create pests. We only need to destroy their habitats and their ways of life, and we will immediately trigger uncontrollable disasters.

As the vegetation continued to increase in this place, I understood that for every species of grass, orchid, tree, or any different plant species, one could observe new insect species. Various animals appeared as plant devourers. For example:

- Small beetles commonly called ladybugs ate the shoots of specific trees.
- Butterfly caterpillars fed on various types of plants.

- New species of orthopterans and cicadas fed on vegetation shoots, eating flower buds and new orchid roots.

Apparently, pests were increasing. However, I just had to observe their life dynamics, and I began acquiring a different perspective. The return of regulators—such as birds, frogs, wasps, ants, mantises, and snakes—fueled my hope. I only had to wait a few more years for this young ecosystem to achieve a natural balance.

The Reality of "Pests" and Environmental Damage

The partial or total destruction of ecosystems forces insects, birds, and animals to adopt new ways of life. The most fragile insects die or move away, while the most resistant ones—which comprise only a few species—fight for survival in the area and feed on the crops. These species reproduce abundantly. For this reason, farmers almost never eliminate the so-called crop pests by applying insecticides; the pests simply become more resistant, while the chemicals kill only the most fragile insects.

During this work of recording and observing the fauna's life dynamics, I realized that by causing progressive damage to nature, coupled with pollution and monocultures, a myriad of birds, animals, and insects could negatively affect all future crops in the Ecuadorian Amazon. Continuous environmental damage changes their eating habits. Out of ignorance, or in a desperate attempt to save the harvest, farmers will resort to the progressive use of chemicals and poisons against birds and animals. This will gradually trigger the beginning of the end of the Ecuadorian Amazon, merely demonstrating that humanity never learned or understood the life dynamics of nature's beings.

I remember that assimilating the insects' way of life was not part of my initial planning; however, this activity became one of the foundational elements of my work. It awakened a passion in me for their beauty and their ways of life. Furthermore, they represent an element that we must not separate or ignore, because ecosystems will never be complete if they lack them.

Emotional Connection and Observation

This represents one of the fabulous experiences I had merely imagined at the beginning. Initially, it created worry, but this later turned into curiosity and ultimately, immense joy. This process allowed me to learn many things I had previously ignored. I remember that while doing this activity, tears welled up in my eyes on several occasions—undeniably out of joy upon realizing that new species chose this place to live. I felt the exact same way when I encountered species I had never met before.

At first, observing the biological activity felt like a pastime and an adventure; however, little by little, I began to understand its profound importance for life. Their presence tells us how much we have restored or conserved; their absence or extinction shows us how much we have destroyed the environment. Loving nature means respecting the life forms of every being within it. Knowing their functions and admiring the intelligence and beauty of each creature will motivate us to maintain this respect, encouraging us to lift our feet and avoid stepping on them as we walk.

Every night in this small forest, I watched new visitors seeking refuge. Their numbers increased every year. The arrivals boasted beautiful colors, designs, and mimicry. Each one became an indicator of their sensitivity to environmental destruction and the healthy conditions the forest was progressively acquiring. The most beautiful insects, birds, and animals cannot exist unless the forest is also beautiful and exhibits an advanced level of health. These conditions remain essential for highly sensitive animals to inhabit the space.

Ecosystem Recovery and the Spider Family

I could not understand where more and more little animals and birds were coming from, but I learned that every single plant species brought new surprises to help balance this nature. Every creature becomes useful when ecosystems develop and strengthen their health. I also began to understand that this methodology for rebuilding ecosystems actually worked. We must recognize that nature might naturally take hundreds of years to achieve the results I obtained in this place over forty years, depending on the sensitivity and threat level of each species. For this reason, we must manually intervene in and support destroyed spaces to achieve their rapid recovery.

Traditionally, experts have compiled lists after conducting studies and research. Perhaps entire libraries exist full of books containing the names of fauna species in danger of extinction or already completely extinct. Even so, the threat of extinction for flora and fauna never ceases. As long as vegetation continues to disappear, imminent threats will surely increase. If we force them out, a large part of this majestic fauna—whose return we eagerly await—could turn into pests for crops and the vegetation itself.

We achieve pest control by changing the eating habits of these beings. We accomplish this by rebuilding ecosystems. Just as we turn them into pests, we can return them to their original ways of life.

Observation represents the most important habit for understanding and learning the perfection of life in nature. In these processes of rebuilding the ecosystem, we achieved the progressive recovery of the trophic chain. I observed that after approximately fifteen years, the creatures we considered pests changed their life habits:

- Caterpillars ate vegetation only at specific times; they also transformed into natural pruners that aided the normal growth of the trees.
- Crickets and cicadas fed on small frogs, caterpillars, and baby spiders.

They were never pests. We gave them that label because, after we destroyed their habitats, they ate our crops. We must understand that they were merely trying to survive.

If we do not surround ourselves with nature, every insect or spider will seem like a problem to our way of life. Instead of looking superficially and labeling them as pests, if we work to restore the ideal spaces they need to inhabit, they will all fulfill their functions. They will provide immense benefits for nature, crops, and human life. When insects cause damage, other insects will appear to regulate their dispersion, and all of them will become part of the balance between nature and humanity.

After twenty-five years, spiders became the most numerous family to arrive in this ecosystem. Over four hundred different species progressively arrived. Each possessed its own way of life. Some species inhabited very specific places, and their homes served as highly aesthetic architectural examples—true fortresses designed to protect themselves and trap their food. Their webs featured complex and varied designs meant to catch specific insects. Because of their ways of life and their intelligence, spiders are incredibly important and merit our thorough observation and study.

Biological Conclusion: "Pests and predators do not exist; only regulators exist."

Photographic Records as Indicators of Forest Health

The following photographs represent a small percentage of the thousands of individuals I recorded as they successively returned while nature developed. Each creature became a potential indicator of the health the forest was acquiring.

Their presence also indicates the degree of sensitivity each species maintains regarding environmental destruction. We observe this based on how long they take to return; as the forest progressively improves its health, each species eventually finds the right conditions it needs to live.

Confirming the Methodology

Confirming the results took many years: the methodology I applied to rebuild ecosystems proved entirely effective. I spent decades waiting for the return of highly important beings that had previously disappeared from this area. I waited for more than forty years, and they finally began to arrive.

For some, their stay remained temporary. The land's conditions—namely its limited size and the surrounding external impacts—do not always favor their permanent residence. Even so, their visit held a profound meaning. Although their future remains uncertain, their presence provided a clear sign that we can restore the earth to the conditions of an almost original paradise.

Chronological Photographic Record of the Return of Fauna

I maintained the photographic record for thirty-four consecutive years (1980-2013), starting with the relevant insects that existed at the beginning, as well as those that arrived sequentially at this place. I performed the majority of this work at night, on various occasions staying out until three in the morning. This activity allowed me to understand their functions, changes in their eating habits, and the degree of sensitivity each species showed toward environmental destruction.

As the years passed, finding spiders eating frogs, crickets eating caterpillars, and snakes eating snakes deeply excited me. I even found animals that died because they ate another individual by mistaking their food. I also recorded anteaters, coatis, otters, weasels, agoutis, pacas, armadillos, and many more. Through this observation, I could understand the approximate number of years it would take to establish ideal habitats for each species to live, alongside the natural order of the trophic chain's formation.

A Passion for Nocturnal Discovery

The nocturnal walks became a passion; so much so that, in the year 2000, I convinced my family to move and live in the forest. This represented another important achievement because it facilitated the night work. I could go out and return to sleep at any time of the night. This activity was not really a job; for me, it was a passion for discovering something new. Progressively, I encountered different animals, snakes, frogs, insects, and hundreds of different spiders. This labor proved crucial for better understanding and knowing why we must respect the life and spaces of every being, no matter how insignificant they may seem to us.

Passing the Torch

Thirty-four years passed very quickly. By 2013, I felt somewhat tired because the daytime work was heavy, followed by nights without a full sleep. I was losing energy even though I found the task emotional and important. I decided I should no longer continue; leaving this activity was difficult. At night, I would grab the flashlight and, along with the little dog that accompanied me, go into the forest because I felt part of it. I actively stopped taking the camera so I wouldn't stay too long, and that is how I gradually left this activity behind. I confess that I no longer possessed the same energy I had when I started at only twenty-three years of age.

Research Results and Ecosystem Variability

The results of the research at this location never constitute a universal law for all restoration zones, because each region possesses different characteristics.

- **Environmental Context:** For example, if a large swamp exists, we would initially find amphibian specimens, or if a forest sits nearby, it will act as the stronghold for the immediate return of fauna.
- **Constant Dynamics:** However, the general patterns I observed in the return of microfauna and macrofauna—based on their energy and the vegetation planted at the restoration site—may recur in other restoration processes.

If a plot of land slated for restoration already contains vegetation, as is the case with secondary forests, some important fauna species will already inhabit it. It will not even be possible to research the trophic chain's formation, because in those places we would not know the order of their arrival based on their sensitivity. However, conducting inventories of existing fauna at different times remains feasible, taking into account that many individuals will inhabit different levels of the forest, making them difficult to spot.

The Photographic Archive

Below, I include a tiny percentage of the photos of animals I recorded each year. I provide these so you can understand their individual attributes and the progressive return of beings that showcase better designs and colors. Their visual quality serves as an indicator of the sensitivity each creature possesses and the progressive health the forest acquires.

Note on the Complete Archive: The author offers a collection of photographs of the species that progressively returned to inhabit the land from 1980 to 2013 (**more than 608 images**), alongside a list of the bird species that also returned.

You can find these resources and many more on the website: https://jardinbotanicolasorquideas.com.

Fauna as an Indicator of Ecosystem Health

When biological recoveries allow food webs to form, we can determine the success of our results. Each individual fulfills the function for which nature created it; each indicates its sensitivity, as well as the forest's conditions and health.

The presence of several spider species actively reduces vegetation damage problems. In a young ecosystem, insects will always eat some plants, but they will no longer kill them. Over the years, you will observe fewer plants broken by insect activity. After approximately forty years, striking ecosystems will form in the forest canopy, giving countless microfauna the opportunity to inhabit those places; we will then observe much less plant deterioration in the lower part of the forest.

Researching Food Chain Dynamics

In a primary forest, knowing the dynamics of the food chain's formation is impossible because everything there works perfectly, and the canopy hosts the vast majority of microfauna. Places where we rebuild ecosystems offer the only possibility to conduct this research. We start without vegetation in these locations, or the vegetation remains small enough that we can easily see what exists and actively track the progressive arrival of micro and macrofauna.

When the vegetation grows large and dense, sequential records become less exact. We cannot see all the individuals because many of them live exclusively high up in the trees.

The Evolution of the Canopy and Forest Strata

In this forest, after thirty years, I observed bromeliads, ferns, philodendrons, and anthuriums growing naturally on the trunks and in the canopy of many trees. Even orchids sprouted because the wind brought their seeds from very distant places.

Consequently, thousands of insects climbed to occupy their new living spaces. Yet, even at night, I still found countless small animals in the lower part of the forest, each occupying different strata according to their ways of life.

As the years pass and the vegetation develops, more sensitive insects, birds, and animals appear. Each one becomes a fabulous indicator of an ecosystem boasting prominent health and characteristics highly similar to those of ancient forests.

Biological Prophecy: "The end of fauna will indicate the end of humanity."

CULTURE

Regarding the cultural element, my primary activity involved engaging in direct and sporadic conversations with people from different communities. I actively sought every opportunity to find information about the historical presence of vegetation, birds, and animals at the restoration site.

I already possessed some knowledge about plant uses, which I acquired from my mother. When I was a child, she treated various of our health problems using jungle vegetation; she also used some wild species for food. For this reason, as I gathered plants to regenerate this forest, my desire to learn about their medicinal, nutritional, artisanal, or cultural values grew. Consequently, I sought opportunities to converse with people from different Indigenous nationalities, who generously shared their wisdom with me regarding the use of specific vegetation. They held this knowledge because their parents, grandparents, or relatives had transmitted these ancestral secrets to them. I primarily focused my interest on learning the values associated with the exact flora I was planting in the ecosystem.

Because I obtained information so sparsely, this process took many years. However, acquiring this knowledge remained absolutely essential. If people forget the importance of certain plants, they simultaneously lose their interest in conserving them, which directly harms specific flora.

We must emphatically highlight the role that different Indigenous nationalities ancestrally fulfilled within nature. Living with the jungle meant:

- Knowing what to eat.
- Knowing how to heal their diseases.
- Discovering the ideal natural materials to craft their weapons and build their homes.
- Figuring out how to achieve a deep connection with nature.
- Carefully selecting their sacred places and plants.

For this reason, we must actively recover this knowledge and strictly avoid the excessive use of nature. Historically, Indigenous peoples used only what they truly needed and never plundered the land. In other words, they represented an indispensable component in maintaining the jungle's ecological balance.

ACTIONS UNDERTAKEN TO MITIGATE IMPACTS DURING ECOSYSTEM DEVELOPMENT

Ecosystem Deterioration Caused by Nearby Inhabitants

When the forest was approximately fifteen years old (in 1995), the vegetation already looked highly attractive. Neighbors entered without prior authorization to take small trees, using them for firewood or to build their traditional homes. Similarly, they cut down *chonta-pambil* palms (*Iriartea deltoidea*) to extract hearts of palm for food; bird and animal hunters also frequently trespassed.

People still believed they had to kill whatever moved, or that they could use nature with extreme freedom, even though they knew they were entering private property. These external actions caused severe deterioration in this tender, still-developing forest. Therefore, I had to find strategies to counteract these damages without causing bitter conflicts with the people involved.

Outreach, Awareness, and Community Engagement

Due to the depredation this young forest suffered, I began talking directly to the people causing the damage. Some people understood immediately. Others actively resisted and argued that they had the right to take whatever they wanted from nature, regardless of the location.

I sought new strategies to convince these individuals and progressively earned their respect for this nature by involving the entire surrounding community. Next, we held meetings and invited children from the nearby school. We planned dedicated training sessions, and the neighbors finally began to understand the true importance of this vegetation.

Little by little, some families approached us with the intention of joining our project. Through brief talks and training sessions, we successfully stopped the damages, allowing the vegetation to develop with our long-desired success. Furthermore, we gained valuable allies to help us conserve nature.

Opening to Tourism Activities

When universities in Ecuador established degree programs in Tourism, Biology, Environmental Studies, and specifically Ecotourism, I saw a profound opportunity. I wanted to spread the importance of educating people to care for, respect, conserve, and rebuild nature.

Thus, eighteen years after I started creating this ecosystem, we held sporadic meetings with family and friends to gauge whether everyday people or tourists would genuinely want to experience this nature. We chose a name and decided the public would identify it as the Amazonian Flora Rescue Center (CERFA), "LAS ORQUÍDEAS" BOTANICAL GARDEN. Consequently, in December 1998, we placed a sign at the entrance and waited for tourists to arrive and visit this place.

Jardín Botánico Las Orquídeas
CERFA
... Un Mundo Natural

Figure 27 - "Las orquídeas" Unified Logo of the Botanical Garden and the Amazonian Flora Rescue Center

People often arrived and asked what the place was about. When I explained that it was a protected space dedicated to nature, many replied with contempt, stating they did not need to pay to see weeds or bush because they could find that everywhere. Similarly, local authorities and tourism institutions actively opposed letting tourists visit this place.

Three years passed like this. The few people who entered—primarily foreigners—did the advertising and highly recommended visiting the garden. The visitors' positive appreciation created immense pressure, eventually forcing local entities to recognize the value of this natural environment. A few local people formed rare exceptions; they valued and recognized the effort, persistence, and profound importance of rebuilding this natural environment right from the start.

Currently, this project has become an environmental conservation model that others replicate in different locations. People actively visit to inquire about the fundamental steps for creating nature, and many individuals are now reproducing nature on their own lands. Naturally, this serves as highly inspiring work for a society that loves and respects the jungle.

Several people have warmly thanked us for inspiring them; others try to hide the fact that we motivated them, even though they visited our site long before undertaking their own conservation efforts. But the most important outcome is that rebuilding this ecosystem ignited a deep, lasting interest in many people to plant nature.

The outreach we have achieved through tourism activities remains truly impressive. Merely by visiting this place, many people have found the inspiration to dedicate their work to nature. That appreciation always was, and continues to be, our greatest strength, because it empowers us to perpetuate this lifelong mission.

Below is the testimony of a person who visited this place. Thank you, Mr. Alberto Martínez, and apologies for publishing what your heart dictates.

Sent from Yahoo Mail for Android Date: Fri, Mar 16, 2011, at 7:16 AM From: Alberto Martínez sepu66@gmail.com

Hello Omar,

My name is Alberto Martínez, and I want to tell you my story with orchids, which started thanks to meeting you.

I was born and live in Spain. In early 2011, the mountaineering club I belonged to in Seville decided to organize a climbing "expedition" to Ecuador. We planned to divide into three groups—northern, central, and southern mountain ranges—and return after climbing the country's most emblematic peaks. We made good friends there from the Quito Polytechnic University Mountaineering Club; a few years later, they even spent many days at my home in Spain. I returned from Ecuador completely fascinated.

I have loved botany since I was very little. At just 10 years old, I already knew some plants by their Latin names. When people asked me what I wanted to be when I grew up, I answered: "a biologist." Ultimately, I studied computer science, and currently, even though I don't like it, I work in an office. However, plants remain my true passion.

Returning to my trip to Ecuador, we took a long break between volcanoes. My passion for plants practically forced me to visit the gates of the Amazon. Given my limited time, visiting the "Las Orquídeas" Botanical Garden in Puyo presented the best option. My partner at the time, Elvira, and I went there. You guided us, and I will always remember that visit, especially because of how deeply I felt what you transmitted.

Before the trip, I already knew some wild Spanish orchids and some hybrid orchids sold in garden centers, but little else. I never remotely imagined how vast and exciting the world of this plant family could be.

During our visit, Omar, you showed us some orchids. I loved seeing them growing in situ, clinging to the tree trunks. But the real impact hit when you took a botanical magnifying glass out of your pocket and placed it on a trunk that seemingly held only mosses and lichens. When I looked through it, my jaw dropped. I saw a perfectly formed orchid flower measuring less than 2 mm. I could not believe such minuscule orchids existed. That moment shocked me, and from then on, I could not get that image out of my head.

When I returned to Spain, I started devouring information about orchids. The more I read, the more I wanted to know. More things caught my attention, and the subject seemed increasingly fascinating. I bought books, set up a home laboratory to germinate seeds, and started looking for interesting orchids to grow at home. My life changed the day I met you. I haven't stopped studying; I am even studying for a degree in Environmental Sciences at the National Distance Education University (UNED) in Spain.

A few years later, when I felt ready, I started writing a blog about orchids. I didn't write about how to cultivate them or their beauty, but rather about their highly interesting peculiarities and their enormous variability, keeping a strictly scientific and botanical perspective. I wanted to spread the message that these plants are more than just decorative flowers. Today, I have over 2,000 followers. My blog address is www.orquimaniaco.es.

Thanks to this, in 2015, I worked for 8 months as a technician at the Estepona Orchidarium Botanical Garden (here is their website: ***www.orchidariumestepona.com***). I spent 4 months preparing everything from scratch. I received plants from Ecuador, Peru, China, Malaysia, and other countries. I placed the orchids on trunks, recreated their natural settings, and prepared specific soils for each terrestrial species. Everything had to look as close to their real habitats as possible, and the plants had to survive and bloom. After the garden's inauguration, I stayed another 4 months to maintain and care for the plants, create new areas and planting surfaces, and study and resolve initial pest problems. The experience proved unforgettable and enormously enriching; I worked with over 1,200 different species.

Since then, I continue using my home laboratory to grow some wild Spanish orchids in vitro. Fortunately, people invite me to many places to give talks and workshops here in Spain. I still cannot make a living from orchids, but little by little, my dream keeps growing.

I owe all of this to my visit to your wonderful little piece of jungle, which you recovered through decades of work, effort, and body-and-soul dedication. Today I wanted to write to you to thank you for everything. I want you to feel proud because my life completely changed thanks to how you infected me with your passion for your work that day.

Thank you very much, Omar. A big hug. Greetings to Ángela as well, even though I didn't get to meet her.

P.S.: I had thought about writing a handwritten letter because I find them more personal and beautiful. Ultimately, I decided to send an email so you could click the links to some websites.

Alberto Martínez Saldaña

Facebook: *https://www.facebook.com/sepu1* Orquimaniaco

Facebook: *https://www.facebook.com/orquimaniacos/*

ENGAGEMENT STRATEGIES IMPLEMENTED AT THE "LAS ORQUÍDEAS" BOTANICAL GARDEN

Promoting tourism in this region provided the opportunity to invite people to experience this small natural environment. I viewed this as another challenge because I felt I was taking on a great responsibility toward nature and the community; furthermore, I needed the opinions of students and people from other regions. Visitors shared many opinions during their tours, but they always remained encouraging and positive, giving me the incentive to pursue my proposed goals.

We developed engagement strategies with nearby communities as a top priority to earn respect for the project and spark society's interest in recovering small or large environments. We successfully recruited allies interested in restoring their own lands and living surrounded by majestic nature once again. In this way, we are fulfilling another dream: seeing more people actively rebuild, value, and respect the life of the jungle.

As a strategy, I decided that the best way to prevent the extinction of certain species was to entrust the care of numerous plant species to many nature lovers. By scattering saplings across many locations, these caretakers guarantee the survival of this nature and ensure we can harvest seeds in the future to continue propagating them.

For several years now, we have actively collected seeds from this forest. For species that do not yet produce seeds, we acquire them from other sources: sometimes people donate seeds to us, other times vendors sell us important seeds or small tree seedlings, and we also go directly into the jungle to search for them. Regardless of the method, we continually acquire species that hold critical environmental, cultural, and biological importance.

Through these efforts, we plant thousands of seedlings in bags and donate them to various communities, provided they demonstrate a genuine interest in conserving and restoring a jungle. I feel absolutely certain that by dispersing plants across different zones, I discovered the most effective way to guarantee the survival of these species.

Figure 28 - Seedbed

Figure 29 - Nursery

Figure 30 - Seed collection

Figure 31 - Seed collection

Since January 2022, through the management of the HUMANS FOR ABUNDANCE Organization, we received valuable support from TERRAFORMATION, who installed the seed bank for us. This allows us to store some of the seeds of the most endangered species to use later during seasons when we lack them.

Climate change is highly evident, actively threatening nature itself. Consequently, several tree species no longer produce seeds every year as they routinely did in the past; numerous species now produce seeds only every three, four, or five years. The most important flora species recede into increasingly distant places. For this reason, we must scatter plants across different locations so that these living germplasm banks will exist in the future. This strategy facilitates the reproduction of all species, actively guaranteeing the lives of both flora and fauna.

Our planning includes the work of reintroducing countless species currently facing the threat of extinction. We will carry out this activity progressively in the rebuilt forests of different communities, where habitats will form naturally according to the needs of each species.

We will never achieve the conservation or protection of flora and fauna if we keep them in captivity. Only freedom will guarantee the survival of these elements; this also applies to Indigenous nationalities, provided that current governments and organizations correct the misguided actions taken during the processes of colonization and civilization.

The "Las Orquídeas" Botanical Garden stands as living proof that rebuilding ecosystems and educating people to respect different forms of life represent the only way to reverse countless environmental impacts. Correcting many of these damages will take generations, but along the way, the results will inspire hope. In this way, I will know that the seeds we sowed here grew deep roots to support massive trees, which will keep the majestic fauna alive. This site serves as an environmental conservation model and an example for others to reproduce. May humanity scatter this knowledge like fertile seeds across the world.

Figure 32 - Preparing trees for delivery

Figure 33 - Delivery of plants to communities

Local Perception

In the cities, few people reflect on what is truly happening to nature. They routinely go to the market and easily find food; they only have to turn on the taps and the water flows. However, they do not realize that authorities must bring this water from increasingly distant places.

Surely, people can sense that the vegetation is no longer so close, or that seeing certain flora or fauna species has become a luxury. They also notice that pollution taints most of the rivers surrounding us, and that water levels in even the largest rivers fail to reach even 50% of their original flow. Today, we no longer catch native fish for food; instead, we rely entirely on farmed fish raised with commercial feed and chemicals. I also believe people must not forget that hurricane-force winds actively destroy many forests and crops.

I feel certain they experience the cold snaps or intense heatwaves, the unexpected droughts that strike during the rainy season, and the torrential rains that flood towns and crops. They feel the burning sun and watch as we scurry like desperate ants trying to avoid its harsh rays. We must understand that these remain relatively minor threats. They show us exactly where we will end up if we fail to restore the damaged areas in time, because nature does not forgive, and it will always demand that we return what we stole from it.

Final Prophecy: "We will have to return to the earth the exact same nature that we stole from it."

CONCLUSIONS AND RECOMMENDATIONS

If you were born and grew up in the city, perhaps you never had the opportunity to feel nature's embrace; you will find it difficult to understand the environmental disaster humans have caused. If you knew exuberant nature and looked at it with greed, exchanging it for money, you will never know the true value of a living jungle. If you never knew or cared to learn what the jungle looked like just in the 1960s, you will fail to differentiate between a natural paradise and the massive destruction present today.

If you had the luck to breathe the oxygen inside a jungle, to enjoy listening to the melodies and songs of the birds, to feel the flight of butterflies, and to watch insects and animals in complete freedom, only then will you understand what eternal prisons mean for them. If you bathed in and drank water from natural lakes, rivers, or the spout of a bamboo channel, you will realize that we no longer have that luxury. Surely, remembering these wonders will bring you to tears.

The Reality of Civilization and Conservation

We must know history to understand exactly where we stand. This way, we will realize what happened to the abundant nature that existed just six decades ago.

We should assimilate that poorly planned civilization and colonization programs started the ruin of Indigenous nationalities, progressively deteriorated the fauna, and now progressively destroy the flora, placing almost every living being under threat and in danger of extinction. We must recognize that leaders planned these events well to achieve their purposes, because even current governments are progressively establishing a high level of extractive activities that threaten every living being in the Amazon.

People always viewed the Ecuadorian Amazon as a set of options from which they could extract massive sums of money to enrich the State and its populations. I still remember them saying out loud, "all resources are inexhaustible." But we must mention that countless species of flora and fauna face threats or the danger of extinction. This has also affected Indigenous nationalities; consequently, some are on the verge of losing even their language.

Although "civilization" benefited certain nationalities by ending conflicts that decimated many jungle inhabitants—whether because they were warriors or had internal conflicts—civilization certainly did not improve their health, education, or economy. Society still completely ignores them, and they remain unaware of the origin of their problems, which outsiders induced through the deceptive promise of civilization and a better life.

Truly, leaders have promoted the few options available to generate economic activities in Indigenous communities without previously analyzing the impacts these actions could create in the short term.

- Example 1: If an organization promotes the production of seed or fiber crafts, it must strictly establish crops of the plants involved in this activity, utilizing proper methods for planting, harvesting, and reproduction.
- Example 2: Regarding medicinal plants and various wild food products, public institutions and government entities readily support the trade, but they never show interest in knowing how vendors obtain the products for the market. They find it easier to ignore the global environmental problems that people consciously or unconsciously create in exchange for economic returns.

Avoiding these impacts would be ideal, but the institutions that help promote the trade of jungle products do not worry; instead, they rejoice at having found another way to make money while causing growing environmental disasters.

The Ecosystem Collapse and The Rise of "Pests"

The diverse palm trees once formed a defining layer of the Amazonian forest landscape. These landscapes recede further and further away, becoming inaccessible for society to enjoy this natural wonder, which holds immense forest, biological, and cultural importance. Today, seeing some of these species is a luxury. Seeing others has become impossible because they have already disappeared from certain places; in fact, several Amazonian provinces in Ecuador now have forests completely devoid of palms.

Following this disaster, the insects' eating habits will change, turning them into pests. Thus, some will attack the remaining palms and their seeds, while others will attack the stems and bases of banana plants, as well as sugarcane. Insects will pierce fruits and seeds, and birds will even devour the leaves of certain crops.

Furthermore, the annual production of many products will progressively shift away from their usual seasons. A lack of pollinators and climate change will delay production to every two, three, or four years.

Raising society's awareness about the importance of respecting life is very difficult, but if we succeed, the solution becomes clear: we must undertake massive restoration to restore the eating habits of everything we call "pests," thereby providing each species with its traditional way of life.

We possess a diverse array of wild vegetation that we could cultivate for food or to generate economic activities without causing environmental impacts. These plants could become part of sustainability as a reward for conservation. We must actively search for these plants, research their uses and properties, generate agroforestry crops, process various types, and commercialize them. We must demonstrate the importance and value of each plant. Because they come from the Amazon, they possess extraordinary medicinal and nutritional properties that would solve society's health and nutrition needs.

The solutions carry high economic costs and require many years to correct the impacts. However, along the way, we could observe the changes and the hope of seeing the Ecuadorian Amazon in all its splendor. We will make it accessible to all of society, since, right now, we cannot all enjoy these natural landscapes that retreat further and further away.

Zoos: Maximum Security Prisons

Many years ago, we could watch many animals, birds, butterflies, beetles, and much more. They all roamed free. They roared and showed their power by marking the space of their domains, or they opened their wings, taking flight and expressing the song of freedom. People find it increasingly difficult to see countless species because nature is running out.

Currently, the Ecuadorian Amazon hosts many ZOOS. These facilities represent MAXIMUM SECURITY PRISONS surrounded by nature, serving as a justification for the widespread greed that arose over sixty years ago. In this way, we prove that we love prison. We let criminals roam free, while we lock innocent beings up for life just because nature endowed them with beauty—because when they were free, they were perfect.

Now, however, their fur and feathers no longer shine as they did when they lived independently, because we entirely changed their diets. Various individuals also reproduce inside these prisons, and authorities call it a success to have more specimens born in captivity. Ecuador's regulatory authority supports these programs, claiming they protect the birds and animals well. Meanwhile, the few Rescue Centers that run wildlife release programs—managed mostly by foreigners—lack the necessary support for maintenance and research. An animal rights law exists, but because these animals come from the jungle, society strips them of those rights.

Surely, the animals and birds can look out at nature from inside their cages. They can see some of the fruits they ate when they were free, but those fruits remain out of reach because humans force them to eat foods that never belonged in their diets. They look outside their prisons at the trees where they once built their nests or burrows, and they surely know that their imprisonment is also part of the great extractivism machine.

A Proposal for Reform

Rescue Centers must necessarily sit next to Nature Reserves to guarantee the release of fauna and their reintegration into their original ways of life. ZOOS are necessary, but we should rename them Rehabilitation Centers. We should delegate to them the care of fauna that we cannot release due to their physical conditions. Specialists at a Rescue Center should issue this diagnosis. These specialists must possess deep knowledge of the region the animals belong to, to properly assess the possibility of release. Otherwise, they must transfer the animals to a Revitalization Center after sterilizing them.

Enlisting the help of autonomous specialists who love freedom would prove highly beneficial. They could offer their professional assessment by thoroughly analyzing each specimen in the zoos; surely, many individuals would have the luck to regain their freedom. We must not fear, because correcting our mistakes does not destroy an economy. Tourists will continue to visit the zoos, and there they will learn a true history, without us changing or inventing reality to justify the eternal imprisonment of innocent beings.

The Work of Restoration

The progressive threat of flora extinction is both worrying and highly noticeable. Countless species turn into relics, and encountering a single specimen has become a luxury. Simultaneously, we must recover their histories and uses.

Because this is a humid tropical region, people seemingly see vegetation everywhere. Abundant secondary forests exist with vegetation that develops naturally. However, if we closely examine the contents, we find that the original trees and plants from primary forests are no longer there. In most cases, these original species will only grow back if we plant them manually. In this way, we will recover the jungle's intense greenery and the balance of every element, all of which depend on each other to establish themselves as a true ecosystem.

The work remains permanent. For those who undertake a restoration, it becomes the path they commit to walking for a lifetime, but in the end, they manage to see the tremendous results. As the years pass and the desired outcome finally appears, it acts as the prize that soothes the exhaustion of the long journey. I remain convinced that nature is an open book for teaching and learning. It is also palpable that environmental damage continues to spread, making the true splendor of nature feel increasingly distant.

I believe I must describe the experiences I lived and the achievements I reached at the restoration site. I hope to touch the hearts of many people so they can feel for nature what I feel.

Perhaps I have become part of the rebuilt jungle because the city stresses me. Not a day goes by that I do not breathe its healthy oxygen and listen to the trill of freedom emitted by the beings that arrive at this, their refuge.

My worry is immense. Children and young people can no longer see those giant trees or those magnificent beetles we used to play with, racing them to see who won. I miss the flocks of beautifully colored butterflies; when they took flight, we would run among them and feel their wings brushing against us. I miss the giant rivers that prompted our parents to warn us, "Be careful, it's dangerous!" and the abundance of fish we could catch with our bare hands. Enumerating those past wonders would take far too long. I still hold the memories of a natural paradise that is progressively dying and rightfully defending itself by creating threats to human life.

A Final Plea for the Future

The rivers are drying up, and they show no signs of recovery. They do not hold even fifty percent of the water I observed some sixty years ago. Populations pollute many of them with waste, mining companies poison others, and oil massively contaminates the rest. People believe the Amazon is massive; well, it is not. Everything remains tiny in the face of the world's immense environmental destruction. As nature continues to shrink, it will fail to soothe the devastating effects of climate change.

We must remain aware that despite the aggression the jungle endures, it still provides us with sustenance to live. Now we must help it, because the moment will come when it can take no more. As a living being that listens and feels, nature will do whatever is unavoidable to survive.

It has been more than forty years now, but it feels like yesterday when I started creating my great jungle. I realize that for a short human life, many years have passed, but for the jungle, this is very little time. Nature will need thousands of years to recover, but if we actively help it, we accelerate its restoration.

The benefits this small rebuilt forest shares are obvious. When I walk, it provides the shade I need to protect myself from the sun's intense heat. My exhaustion vanishes when I listen to the music the birds and insects create. The breeze moving through the forest creates harmony with the vegetation. I admire the beauty of the birds and their joyful songs, which manifest their freedom. I now feel the greatest reward after the intense work I've done since 1980. Furthermore, it brings me joy that many people breathe the oxygen the wind disperses from this forest.

Today, right here, I can demonstrate that no fortune in the world can emanate the pure air we breathe here. No amount of money can buy the joy of watching more than a hundred and eighty bird species sing and fly. Migratory birds arrive during specific seasons, while others visit permanently. Surprisingly, many of them had completely disappeared from this area many years ago. But the impossible returned to this natural refuge. They all sing with joy because they are free, and the vegetation provides the very elements they need to live.

Sometimes tears fill my eyes when I walk and encounter monkeys who trust that we respect their ways of life. They surely traveled from very far away to settle on this island of nature. These beings jump from tree to tree because they find their food here; they scatter seeds and consider this a proper place to live. Yet, I observe with great sadness that a large part of this will inevitably be lost again. The city has caught up with me, and it will continue its siege, surrounding this forest with footprints of death.

Words cannot express the joy of enjoying nature's peace at night. Its coolness and the songs of the insects serve as a manifestation that demands our understanding and willingness to learn how to interpret what they are telling us. Observing the forest during the day proves deeply inspiring. However, listening to the songs and sounds at night allows us to comprehend that life in nature does not make itself visible to every human eye. I remember that during my continuous nocturnal walks, I encountered important animals such as:

- Otters
- Caimans
- Armadillos
- Coatis
- Anteaters
- Weasels

The "Las Orquídeas" Botanical Garden pioneers a new concept of what traditional Botanical Gardens represent. Here, we essentially manifest the roles of conservation, restoring the land to serve as a model that people can apply in different regions. Surely, the results will differ. Planners might also omit certain elements depending on the specific location or region, as the advantages and difficulties of the restoration process will inevitably vary.

Saving flora and fauna is a labor driven by pure conviction. We will likely never fully reach the finish line, nor will we ever declare our duty completely fulfilled. Along this path, countless indispensable activities and proposals will surely emerge. To execute them, we constantly need allies who share this exact passion. We will draw our final breath with satisfaction, continuously striving to make all of humanity the greatest friends of nature.

Given the current rhythms of destruction, we easily perceive that maintaining crops in the Ecuadorian Amazon will become progressively more difficult. Because of accelerated environmental destruction, many beautiful birds, important animals, and insects will transform into major crop pests. Many people will lay down poison, set traps, and spray insecticides, and the problem will only grow worse. If we continue with this same rhythm of life, finding organic products in the Amazon will soon become incredibly difficult.

The temperature will rise to previously unrecorded levels. Nature used to sustain itself with abundant water; consequently, the decrease in humidity will trigger scorching temperatures. This risk paves a dark path straight toward an Amazonian desert. The only true solution involves rebuilding ecosystems on a massive scale. We can confidently solve every impact, but we urgently need a widespread shift in consciousness and decisive action before time runs out. Considering the sheer scale of destruction, funding these genuine solutions costs a fortune, and we must wait a long time to see the results.

The spaces contaminated by mining and oil spills cause deep concern as well, because their recovery will take many generations. Industries permanently pollute the rivers that feed directly into the Amazon River. Yet, people fish in these rivers for personal consumption and commercial sale, tourists visit them, and locals bathe in them. This toxic water also contaminates the crops. Later, people will wonder why diseases suddenly afflict the jungle's inhabitants and general visitors.

Do the polluters know how to decontaminate?

Absolutely not!

They only care about the fortune they amass at the expense of thousands of lives. This goes far beyond mere aggression against nature; it constitutes a progressive, direct assassination of humanity itself.

If we combine all the environmental projects currently underway in the Amazon, we amount to a mere speck in the middle of this vast destruction. At this point, simply changing our attitudes no longer offers a solution; we must actively battle the powers of destruction.

Sometimes I ask myself, and I want to ask society:

Will we manage to avoid the disaster?

References

Bass, M., M. Finer, C. Jenkins, et al. 2010. "Global Conservation Significance of Ecuador's Yasuní National Park." *PLoS ONE* 5 (1).

Ecuador. 1996. Executive Decree No. 195. Supplement to the Official Register No. 40, October 4.

Ecuador. 1999. Executive Decree No. 505. Official Register No. 118, January 28.

El Diario. 2010. "Registro Oficial" [Official Register]. August. www.eldiario.com.ec.

Global Forest Watch. 2026. "Ecuador." https://www.globalforestwatch.org/map/country/ECU/.

Martinez, Alberto. 2009. "Saludos desde España" [Greetings from Spain]. Email received by Omar Tello, March 12.

Science Panel for the Amazon. 2021. *Amazon Assessment Report.* Chapter 19.

Viteri Díaz, G. 2007. *Reforma Agraria en el Ecuador* [Agrarian Reform in Ecuador]. www.eumed.net/libros/2007b/298/.

Wunderling, N., et al. 2026. "Deforestation-Induced Drying Lowers Amazon Climate Threshold." *Nature*, March 26. https://doi.org/10.1038/s41586-026-10456-0.

APPENDIX A

Cartography of Amazonian Resistance and Defense: A Conceptual Framework for Omar Tello's Testimony

By Jon Dell Jaramillo, PhD

The work *The Man Who Planted His Own Jungle* transcends the memoir genre to establish itself as a fundamental document of Amazonian political ecology and a pioneering restoration methodology. Omar Tello's testimony is not a simple reforestation anecdote; it is a judicial chronicle, an allegory of cultural dependence—as reflected in the domesticated monkey symbol—and a direct confrontation with the power structures that have orchestrated the ecosystem's devastation.

In this sense, the book also belongs to the long Latin American tradition of *testimonio* as a political genre: a singular voice that speaks, as John Beverley (2004) affirms, not only for itself but as a representative of a silenced collective experience. That this silence belongs to the jungle and its historical inhabitants makes Tello's book an exceptional case within this tradition.

This appendix offers a conceptual framework for the reader to situate Tello's struggle within global academic debates. By integrating the author's narratives with concepts of environmental justice, biocentrism, Indigenous epistemology, feminist political ecology, and slow violence, it argues that the response to the Amazonian crisis is inherently cultural, biological, and political, and never merely technical. The conclusion of this appendix proposes a tripartite cartography in which each of these registers—the epistemic, the social, and the environmental—corresponds to a different dimension of the justice that Tello's testimony demands.

1. Biocentrism and Earth Jurisprudence: The Ethical Foundation of Restoration

From the first chapter, Tello's narrative establishes a non-negotiable ethical principle: the Amazon is a living being, possessing rights, that was violated without consent. The author, recalling that humanity should have "asked permission" of the jungle, anticipates and embodies the spirit of Biocentrism, which posits the intrinsic value of life, independent of its utility for human beings. This intuition is not a rhetorical ornament; it forms the ethical backbone of the entire methodology the author develops in the following chapters.

Biocentrism directly opposes the anthropocentrism that has historically justified "forest harvesting" and extractivism. The legal genealogy of this opposition traces back to a precise point of origin: Christopher Stone, in his seminal *Should Trees Have Standing?* (1972), argued that natural objects—forests, oceans, and rivers should possess their own legal rights, independent of their instrumental value to humans. Although academics considered Stone eccentric at the time, his thesis has become the legal foundation for rights of nature movements worldwide. Ecuador's 2008 Constitution, which recognizes Nature (*Pachamama*) as a subject of legal rights, serves as its direct heir and the implicit legal framework for Tello's entire action.

Cormac Cullinan's (2011) concept of *Wild Law* provides the jurisprudential complement: human laws must remain consistent with Earth Jurisprudence, understood as the planet's essential functioning. By restoring the soil with sawdust humus (Chapter III), returning the original flora, and creating the conditions for the free monkey's return (Chapter II), the author acts not only as a botanist but as an ecological jurist who restores the ecosystem's violated rights. Decades before the category fully existed, his practice anticipates what Anna Tsing (2015) would call an ecology of cohabitation: the idea that humans do not manage nature but rather cohabit alongside it in relationships of mutual dependence and implication.

Sacred Ecology (Berkes 2012; de la Cadena 2015) also nourishes this ethical perspective. Tello's childhood feeling that the jungle "must be treated as something sacred" goes beyond simple nostalgia; it recognizes the spiritual function nature has historically held as a conservation mechanism. In Amazonian environmentalism, the biological and the sacred constitute a single ontological sphere. The author's prophecy, "The jungle will cry, but with its tears, it will flood the aggressors," resonates with the Amerindian ontology of cosmic justice that Eduardo Viveiros de Castro (1998) termed Perspectivism: the understanding that non-human beings possess their own perspectives on the world, which are just as legitimate as human ones. The restorer thus becomes an officiant who repairs physical and spiritual damage, restoring not only the flora but the web of intertwined perspectives that constitute the jungle's life.

2. The Critique of the Extraction Economy and the "Great Debt"

Tello's testimony dismantles the "progress" narrative that drove the colonization of the Ecuadorian Amazon starting in the 1960s and 1970s. His account details how state policy, operating through institutions like IERAC, served as an instrument of economic and ecological dispossession, creating what the author calls the "great debt"—a debt that is not only financial but biological, cultural, and historical.

2.1. The Bureaucratic Instrument of Destruction

The Agrarian Reform Law (1964/1973) and IERAC did not represent acts of development; they constituted acts of war against the forest. In Tello's account, IERAC required colonizing families to clear as much as eighty percent of the forest in order to demonstrate productive use and secure ownership. He interprets this policy as transforming the forest from a common good into a commodity and directly linking legal ownership to ecological destruction. This resulted in the widespread adoption of pasture-based cattle ranching and monocultures—such as gramalote grass and naranjilla—processes that Tello documents as the "first two failed lessons" nature taught him (Chapter I).

2.2. Dependency Theory and Unequal Exchange

The promise of fortune for colonizing peasant farmers proved to be a chimera. Tello is emphatic: "I do not remember, nor do I know a single farmer who made a fortune or lived comfortably with this activity" (Chapter I). This observation can be read through the lens of the Unequal Exchange theory of political economy (Frank 1967; Galeano 1971). The peasant farmer performs the most arduous labor—felling centuries-old trees with an ax, carrying the wood on his shoulder—only to receive a ridiculous payment. Merchants operating processing factories and large export companies capture the real fortune (Chapter I). In this model, external forces extract ecological value and human effort, transferring them to the metropolis or the elite. This leaves the local producer mired in debt and impoverishment, despite living in a territory of "unparalleled wealth" (Acosta and Cajas-Guijarro 2020).

2.3. From Conflict to Land Concentration

The Agrarian Reform, theoretically designed to distribute land, paradoxically ended up exacerbating inequality. Tello documents how exhausted and indebted peasant farmers sold their plots, resulting in the formation of new landowners and the reconcentration of land (Ladejinsky and Walinsky 1977; Kay and Akram-Lodhi 2021; Levien et al. 2019). Amazonian development policy thus transformed into a tool that facilitated both the underlying oil extractivism (Valdivia 2022; Kimerling et al. 1991) and the consolidation of the cattle latifundia that today threatens the remaining biodiversity.

3. Slow Violence, Epistemicide, and Bureaucratic Resistance

Tello's work does not merely critique historical colonization; it rigorously examines the mechanisms of destruction that persist in the present, ranging from urban pollution to state persecution against those attempting to restore the land.

3.1. The Gradual Disaster: Slow Violence

The destruction of the Amazon represents, to a large extent, a phenomenon of slow violence in the sense Rob Nixon (2011) gives the term: a violence that unfolds dispersed across time and space, that is fundamentally narrative, and that remains politically invisible precisely because of its slowness and dispersion. This violence manifests in rivers losing fifty percent of their flow, the scarcity of native fish, and deeply eroded soils. It is a disaster unfolding over decades, difficult to capture in media headlines, yet

Tello records it with testimonial precision: the garbage dumped daily into rivers (Chapter I) and the collective forgetting regarding the fact that "resources are [not] inexhaustible" (Chapter III).

Elizabeth Povinelli (2016) expands this reading with the concept of *geontopower*: the power exercised through distinguishing between what counts as "life" and what society classifies as mere "geontology"—earth, rock, soil, water. Under the logic of geontopower, the felled jungle is not a murdered living being but an exploited resource; the eroded soil is not a wound but an externality. Tello's book actively counters this logic by treating humus, fermented sawdust, and Amazonian soil as organisms possessing history, agency, and rights of restitution. His restoration work operates, among other things, as an act of resistance against the geontopower that has made damage to the non-living politically invisible.

3.2. Epistemicide and Cultural Defense

Extractivism's most profound attack is not solely environmental but epistemic: the dispossession of ancestral knowledge, which Tello calls "the true science" of Indigenous nationalities. Colonization and evangelization brought disease and material dependence, but furthermore, as the author argues, "tearing that link [with nature] away from them was tearing away their soul" (Chapter II). Outsiders stripped the wise elders—the living memories and community guides—of their knowledge, subsequently selling it as an exotic commodity in the academic and pharmaceutical markets of the Global North (Hamilton 2006; Dutfield 2009; Robinson 2010).

Boaventura de Sousa Santos (2014) terms this systematic destruction of non-Western knowledge forms *epistemicide*, identifying it as a structural condition of modern colonialism. The domesticated monkey allegory (Chapter II) brilliantly summarizes this tragedy: the monkey, stripped of his freedom and his knowledge of the jungle, becomes a dependent and destructive being, whose only hope resides in the reconstruction of the original habitat. This mirrors the exact crisis facing Indigenous nationalities pushed to the urban periphery and forced to abandon their worldview and science. In this sense, Tello's restoration functions not just as a botanical act but as an act of epistemic recovery: restoring the jungle means restoring the context in which certain types of knowledge remain possible.

3.3. The Struggle Against Bureaucratic Control

Perhaps the most revealing passage concerning power structures is the twenty-year state persecution narrated in Chapter I (Part 6). When the author attempts to rescue seeds and plants from logged forests, the environmental bureaucracy—an institution created to protect nature—threatens him with prison and fines, operating under the dogma that he "must let them rot, and you should not care if any species goes extinct." Tello presents this conflict as a paradox of environmental governance: in his experience, regulatory authority was directed against his efforts to rescue and propagate threatened plants while much larger forms of environmental destruction continued around him.

4. Feminist Political Ecology and the Territorial Dimension of the Body

One dimension that the literature on ecological restoration in the Amazon has been slow to recognize is the concept feminist political ecology articulates as *body-territory*: the idea that women's bodies and community bodies serve as the first colonized territory, making their defense and the defense of the forest inseparable. Rocío Silva Santisteban (2017) has developed this framework within the Andean and Amazonian context, arguing that extractivist violence inscribes itself simultaneously on the land and on the bodies of those who inhabit it, particularly Indigenous women.

Although Tello's testimony does not explicitly thematize gender, his narrative contains moments that strongly resonate with this framework. The maternal figure of the jungle—the forest that cries, nurtures, and possesses a soul—and the description of wise elders as the "guides" of the community suggest an economy of care that feminist political ecology recognizes as central to the ecosystem's social reproduction. Gladys Tzul Tzul (2019), writing from the perspective of Indigenous communal government systems, has demonstrated that women frequently sustain the networks of reciprocity and collective management that make territorial resistance possible. The "seedling dispersion" strategy that Tello articulates for conservation—donating to communities, conducting environmental education, creating the botanical garden— can be read alongside this logic of distributed community care.

Incorporating this framework into the book's analysis is not a gesture of political correctness; it serves as a crucial method to complete the cartography. We cannot solve the Amazonian crisis if we view it solely as a problem of resource management or ecological engineering. As feminist political ecology insists, resolving this crisis requires a total transformation of the power relations that organize who cares for the land, who makes the decisions, and who ultimately bears the cost of the disasters.

5. The Science of Restoration: Fragmentation, Thresholds, and Tello's Methodology

The restoration methodology Tello intuitively developed over decades finds robust support and depth in contemporary scientific ecology. Understanding this dialogue—between local practice and global science—proves fundamental to appreciating the originality and scope of his contribution.

5.1. Forest Fragmentation and Collapse Thresholds

For decades, William Laurance and his collaborators have documented the complex ecological consequences of Amazonian forest fragmentation, including edge effects, changes in species abundance and composition, altered forest dynamics, and the important influence of the surrounding landscape matrix on connectivity and ecosystem functioning (Laurance et al. 2002). Thomas Lovejoy and collaborators, through the long-running Biological Dynamics of Forest Fragments Project in the Brazilian Amazon, helped establish an experimental framework for understanding how fragment size, isolation, edges, and surrounding vegetation affect tropical forest communities (Lovejoy and Bierregaard 1990).

Tello's work takes on a new dimension when read alongside these findings. His 6.75 hectares can be understood as a small restoration node within an increasingly fragmented landscape. His use of clustered plantings and strategically distributed restoration nuclei bears a notable resemblance to nucleation approaches in restoration ecology, which seek to establish patches from which ecological interactions and vegetation recovery can expand into surrounding degraded areas (Reis et al. 2010). Rather than demonstrating an exact equivalence between his practice and fragmentation theory, this parallel shows how Tello's locally developed methodology addresses some of the same problems of connectivity, dispersal, and ecological interaction identified in the scientific literature.

5.2. The Soil as Subject: Beyond the Substrate

The intense attention Tello dedicates to Amazonian soil—its thinness, fragility, and slow formation—reflects an understanding that soil science has struggled to convey to public policy. Carl Jordan (1985) documented that tropical forest nutrient cycling places substantial importance on nutrients held in vegetation and rapidly recycled organic matter, while many highly weathered Amazonian soils contain relatively limited readily available nutrients. This helps explain why clearing forest can produce only temporary increases in nutrient availability before nutrient losses and declining soil fertility become significant under many forms of cultivation. Tello's creative solution—constructing humus layers using sawdust and sugarcane bagasse—provides a practical response to this structural problem.

From Povinelli's (2016) perspective, this attention to the soil also constitutes a political act: treating humus not as an inert substrate but as an entity possessing history and restitution needs actively resists the geontopower logic that turns the earth into a mere resource. Tello does not restore *on* the soil; he restores the soil itself, recognizing that without addressing this foundational dimension, reforestation remains entirely cosmetic.

6. Testimonio as a Form of Knowledge and Political Action

The Man Who Planted His Own Jungle firmly belongs to the Latin American *testimonio* genre, a literary and political form that Beverley (2004) defines as a first-person narrative of a lived experience demanding social urgency and non-fictional veracity.

Unlike bourgeois autobiography, which centers on individual singularity, the *testimonio* speaks from and toward a collective. Tello does not narrate his personal story; he narrates the history of a silenced forest, a silenced community, and a silenced ecosystem.

This generic dimension carries profound epistemological consequences. George Yúdice (1991) noted that the *testimonio* does not merely serve as a data source for academia; it functions as a political intervention producing *situated knowledge*—knowledge generated from below, from the physical body, from the direct experience of destruction and restoration. In this sense, Tello's book does not merely illustrate concepts associated with Nixon, de Sousa Santos, or Cullinan; his lived experience provides a situated Amazonian case through which those theoretical frameworks can be read and tested against the realities they seek to describe. From this perspective, "slow violence" is not merely an academic concept applied retrospectively to the Ecuadorian Amazon: Tello's testimony records the kind of cumulative, dispersed environmental harm that Nixon's term helps us name.

Recognizing Tello's testimony as a legitimate form of ecological and political knowledge constitutes, in itself, an act of epistemic justice. It requires academia to accept that knowledge produced from local practice—from the body that works the land and suffers its degradation—holds an authority that academic models cannot replace, even though they can accompany and amplify it. This appendix aims to do precisely that: to serve as a theoretical accompaniment to a body of knowledge that already possesses its own integrity and epistemic value within the pages that precede it.

7. Who Defends the Jungle? A Typology of Restorative Actors

Tello's methodology—based on the integration of Forest, Biological, and Cultural elements—offers a practical and poetic answer to the question of who defends the jungle, articulating a typology of actors ranging from the individual to non-human agents.

7.1. The Integral Restorer (The New Botanist)

The author embodies the role of the Integral Restorer, taking on a task that transcends traditional reforestation, which often centers on planting trees for later timber use. Tello's practice bears a notable resemblance to approaches described within restoration ecology as ecological nucleation and assisted succession: he establishes "islands" of palms and key species and proposes strategically distributed restoration nuclei intended to facilitate the movement of seeds, pollinators, and fauna across degraded spaces (Reis et al. 2010). The resemblance places Tello's independently developed practice in productive dialogue with contemporary restoration theory without requiring that the two be treated as identical.

His approach also addresses the physical limitations he observed in degraded soils through nutrient recycling. Tello describes creating organic soil layers from sawdust, sugarcane bagasse, and animal waste, a practice he understood as accelerating the rebuilding of conditions necessary for forest development. Finally, as a Germplasm Guardian, he describes the concentration and propagation of diverse plant species across his 6.75 hectares as a living germplasm bank. In his account, this strategy provides a practical response to increasingly irregular seed production and improves access to plant material for future propagation and restoration.

7.2. The Biological Defender (Fauna as Indicator)

In Tello's vision, fauna transcends its role as a passive victim, becoming both an ecological participant and an indicator of environmental recovery. His central hypothesis, that "planting the lost flora will bring back the lost fauna," finds substantial support in the longitudinal observations he documents at Las Orquídeas. Tello reports the return of more than 180 bird species, together with fauna such as otters, caimans, and anteaters. Particularly significant within his methodology is the reappearance of glass frogs after approximately thirty years, an observation he interprets as evidence that the reconstructed forest had developed increasingly favorable ecological conditions. These observations do not by themselves constitute controlled scientific verification of the methodology, but they can be placed in productive dialogue with ecological research on habitat complexity, connectivity, and species recovery.

Tello extends this ecological interpretation through what might be called his critique of the "Myth of Pests," expressed in the maxim: "Pests and predators do not exist; only regulators exist." Drawing on his observations at the restoration site, he describes insects that initially damaged planted vegetation but later became incorporated into increasingly complex feeding relationships as plant diversity and other fauna returned. He interprets this change as evidence that ecological imbalance, rather than the inherent nature of particular organisms, can produce what humans classify as pest behavior. His conclusion resonates with principles found in agroecology and ecological pest management, although Tello's formulation remains his own experiential and ethical interpretation of those processes.

7.3. The Ally in Resistance (Seed Dispersal)

The book proposes a defense model that transcends the author's individual figure, articulating a plan based firmly on the principles of delegation and humility. This vision materializes in conscious ecotourism: the opening of the "Las Orquídeas" Botanical Garden does not pursue economic gain, but functions as a critical mechanism for environmental education and community engagement. Alberto Martínez's testimony (Chapter III) illustrates how visiting the restored jungle becomes a transformative act, capable of igniting passion and action on a global scale. The dispersion strategy delegates conservation to the communities through the donation of thousands of seedlings, effectively turning humans into conscious seed dispersers.

However, it is precisely here that nature demonstrates its absolute superiority with a dose of dark humor. While the author and the communities serve as key allies, they are certainly not the most efficient. The bat, the Amazon's true nocturnal gardener, disperses seeds across remarkable distances in a single night, turning its biological efficiency into a subtle mockery of the slowness of bureaucrats who waste years debating permits in the capital. Bees and other insects ensure genetic diversity without ever consulting a procedure manual, highlighting the deep irony of a State that persecutes the author for "stealing" seeds while simultaneously permitting the use of agrochemicals that annihilate these unpaid agents. Monkeys and other frugivores act as the jungle's transporters, carrying seeds into places human restorers might never reach and giving biological expression to one of Tello's central convictions: biodiversity flourishes through movement, interaction, and freedom rather than confinement.

Ultimately, the Integral Restorer and his community allies provide the necessary framework, while the forest's regenerative processes increasingly depend on the continuing activity of the non-human beings that recolonize it. Their ecological functions reveal the limits of what human restoration efforts alone can reproduce. This conclusion also serves as a response to geontopower: the agents of restoration are not only human, but encompass the wider web of life that Tello has worked to reconstitute.

Conclusion: Restoration as an Act of Tripartite Justice

The Man Who Planted His Own Jungle ultimately presents restoration as an act operating simultaneously across three registers of justice: epistemic justice, social justice, and environmental justice. These three dimensions are not simply themes the book addresses; they represent the three structural registers where the damage operates—the damage Tello dedicated his life to repairing—and each corresponds directly to a section of the conceptual framework this appendix has constructed.

Epistemic justice—developed in sections 1, 3.2, and 6—demands the recognition that the ecological knowledge held by Indigenous nationalities and practical restorers like Tello constitutes legitimate and urgent knowledge, not a mere ethnographic curiosity. Epistemicide did not end with colonization; it continues in every environmental policy that ignores local knowledge, and in every bureaucracy that threatens seed rescuers with prison. The only adequate response is what de Sousa Santos calls an *ecology of knowledges*: not the subordination of Indigenous and practical knowledge to Western scientific knowledge, but their horizontal dialogue.

Social justice—developed in sections 2 and 4—demands the recognition that the Amazonian crisis remains inseparable from the inequality structures that turned the jungle into a commodity and peasant farmers into instruments of their own destruction. Unequal exchange, land concentration, and oil extractivism are not historical accidents; they represent the result of deliberate policies that distributed the costs of destruction among the poorest and the profits among the richest. True Amazonian social justice requires not only rights of nature laws, but material redistribution: land, water, seeds, and knowledge.

Environmental justice—developed in sections 3.1, 5, and 7—demands the recognition that the slow violence of ecological deterioration functions as a form of political violence that disproportionately affects those who contributed least to it. Rivers losing their flow, soils losing their humus, forests crossing thresholds of irreversible collapse: global extractivism inflicts these wounds on local communities who bear them in silence. Tello's restoration stands as the most direct possible response to this injustice: it is not an abstract demand, but a concrete action, restoring humus millimeter by millimeter, returning flora species by species.

If global society truly recognizes the historical and political "great debt" it owes the Amazon (Kimerling et al. 1991), the only way to repay it is not with empty hope, but with concrete actions like Tello's model. His testimony presents the restoration of three interconnected elements—the forest, the fauna, and the culture—as more than the planting of flora: within Tello's vision, it becomes a practice aimed at restoring the conditions in which living beings can recover habitat, autonomy, and dignity. The final question the book leaves us is not whether we can avoid disaster, but whether we possess the ethical will and political courage to take on the mission of "reintroducing to the earth the exact nature we stole from it during our lives."

References

Acosta, Alberto, and John Cajas-Guijarro. 2020. "Ghosts, Pluriverse and Hopes: From 'Development' to Post-Development." In *The Routledge Handbook to Global Political Economy*, edited by Ernesto Vivares, 292–308. Routledge.

Berkes, Fikret. 2012. *Sacred Ecology*. Taylor & Francis.

Beverley, John. 2004. *Testimonio: On the Politics of Truth*. University of Minnesota Press.
Cullinan, Cormac. 2011. *Wild Law: A Manifesto for Earth Justice*. 2nd ed. Chelsea Green Publishing.
de la Cadena, Marisol. 2015. *Earth Beings: Ecologies of Practice across Andean Worlds*. Duke University Press.
de Sousa Santos, Boaventura. 2014. *Epistemologies of the South: Justice Against Epistemicide*. Routledge.
Dutfield, Graham. 2009. "Protecting the Rights of Indigenous Peoples: Can Prior Informed Consent Help?" In *Indigenous Peoples, Consent and Benefit Sharing*, edited by Doris Schroeder et al., 53–67. Springer.
Frank, Andre Gunder. 1967. *Capitalism and Underdevelopment in Latin America*. Monthly Review Press.
Galeano, Eduardo. 1971. *Las venas abiertas de América Latina* [Open Veins of Latin America]. Siglo Veintiuno.
Hamilton, Chris. 2006. "Biodiversity, Biopiracy and Benefits: What Allegations of Biopiracy Tell Us About Intellectual Property." *Developing World Bioethics* 6 (3): 158–73.
Jordan, Carl F. 1985. *Nutrient Cycling in Tropical Forest Ecosystems*. University of Michigan International.
Kay, C., and A. H. Akram-Lodhi. 2021. "The Diversity of Classical Agrarian Marxism." In *Handbook of Critical Agrarian Studies*, 25–33. Edward Elgar Publishing.
Kimerling, Judith, et al. 1991. *Amazon Crude*. Natural Resources Defense Council.
Ladejinsky, Wolf Isaac, and Louis Joseph Walinsky. 1977. *Agrarian Reform as Unfinished Business*. World Bank.
Laurance, William F., et al. 2002. "Ecosystem Decay of Amazonian Forest Fragments: A 22-Year Investigation." *Conservation Biology* 16 (3): 605–18.
Levien, Michael, et al., eds. 2019. *Agrarian Marxism*. Routledge.
Lovejoy, Thomas E., and Richard O. Bierregaard. 1990. "Central Amazonian Forests and the Minimum Critical Size of Ecosystems Project." In *Four Neotropical Rainforests*, edited by Alwyn Gentry, 60–71. Yale University Press.
Nixon, Rob. 2011. *Slow Violence and the Environmentalism of the Poor*. Harvard University Press.
Povinelli, Elizabeth A. 2016. *Geontologies: A Requiem to Late Liberalism*. Duke University Press.
Reis, Ademir, et al. 2010. "Nucleation in Tropical Ecological Restoration." *Scientia Agricola* 67 (2): 244–50.
Robinson, Daniel. 2010. *Confronting Biopiracy: Challenges, Cases and International Debates*. Routledge.
Silva Santisteban, Rocío. 2017. *El factor Quispe: nacionalismo, raza y ciudadanía*. Sur Casa de Estudios del Socialismo.
Stone, Christopher D. 1972. *Should Trees Have Standing? Toward Legal Rights for Natural Objects*. William Kaufmann.
Tsing, Anna Lowenhaupt. 2015. *The Mushroom at the End of the World: On the Possibility of Life in Capitalist Ruins*. Princeton University Press.

Tzul Tzul, Gladys. 2019. *Sistemas de gobierno comunal indígena: mujeres y tramas de parentesco en Chuimeq'ena'*. Sociedad Comunitaria de Estudios Estratégicos.
Valdivia, Gabriela. 2022. "Of Extractivist Wounds and Healing in the Americas." *Latin American Research Review* 57 (4): 960–71.
Viveiros de Castro, Eduardo. 1998. "Cosmological Deixis and Amerindian Perspectivism." *Journal of the Royal Anthropological Institute* 4 (3): 469–88.
Yúdice, George. 1991. "Testimonio and Postmodernism." *Latin American Perspectives* 18 (3): 15–31.

GLOSSARY

This glossary compiles the scientific, ecological, and cultural terms used throughout the text. It organizes them into three sections: Flora and Fauna Species, Ecological and Technical Terms, and Cultural and Historical Terms. The entries have been alphabetized by their English translations, with the original Spanish terms provided in parentheses.

I. Flora and Fauna Species

- **American mahogany (*Ahuano*) — *Swietenia macrophylla* (MELIACEAE):** A highly valuable timber tree also known as American mahogany. Coveted for its durable, high-quality reddish wood. Currently threatened with extinction due to intensive extraction. Indigenous communities ancestrally used it to build canoes.
- **Araceae (*Araceae*) — Botanical family:** A family of tropical plants that includes herbaceous and epiphytic species characteristic of Amazonian forests. Its representatives include anthuriums and *Philodendron*. They act as important natural colonizers in ecosystems undergoing restoration.
- **Balsam (*Bálsamo*) — *Myroxylon balsamum* (FABACEAE):** An Amazonian timber tree, prized for its aromatic resin used for medicinal and religious purposes. Its wood boasts high durability. It figures among the species intensively exploited during the colonization era.
- **Bitter cane (*Caña agria*) — *Costus sp.* (COSTACEAE):** A medicinal herbaceous plant widely recognized in the Ecuadorian Amazon. Indigenous communities use its leaves and stems to treat various diseases. It represents an example of the ancestral Amazonian pharmacopoeia.
- **Cassava / Yuca (*Yuca*) — *Manihot spp.* (EUPHORBIACEAE):** An ancestral tuber crop, fundamental for feeding Indigenous peoples and Amazonian colonizers. It grows well in poor soils, making it an essential subsistence crop in the Amazon.
- **Cat's claw (*Uña de gato*) — *Uncaria tomentosa* (RUBIACEAE):** A medicinal liana of recognized importance in the Amazonian Indigenous pharmacopoeia. Traditional

medicine uses it as an anti-inflammatory and to treat various diseases. Uncontrolled commercial harvesting has placed it under pressure in many Amazon sectors.

- **Cedar (*Cedro*) — *Cedrela odorata* (MELIACEAE):** A tree with fragrant and light wood, highly valued in the Amazon region. Indigenous communities used it to build canoes. Its massive extraction has driven it to the brink of extinction in many sectors of the Ecuadorian Amazon.
- **Chonta / Chonta pambil (*Chonta / Chonta pambil*) — *Iriartea deltoidea* (ARECACEAE):** An emblematic palm of the Ecuadorian Amazon. Its extremely hard and incorruptible wood requires over two hundred years to reach maximum quality. It provides a source of food, tools, and building material for Indigenous communities. During the restoration process of the "Las Orquídeas" Botanical Garden, centuries-old specimens of this species proved decisive for seed collection.
- **Chuncho (*Chuncho*) — *Cedrelinga cateniformis* (FABACEAE):** A large timber tree, one of the tallest in the Amazon. Loggers heavily exploited its wood for construction and canoe making. It appears on threatened species lists due to historical deforestation.
- **Costaceae (*Costaceae*) — Botanical family:** A family of tropical herbaceous plants, represented in the Amazon mainly by the genus *Costus* (*caña agria*). Its members appear naturally during ecosystem restoration, contributing to the understory's equilibrium.
- **Gramalote grass (*Gramalote*) — *Axonopus scoparius* (POACEAE):** A large grass massively used in the Ecuadorian Amazon for cattle ranching. Its implementation required clearing thousands of hectares of primary forest during colonization processes, making it one of the main symbols of the region's forest destruction.
- **Guambula (*Guambula*) — *Minquartia guianensis* (OLACACEAE):** An Amazonian timber tree, known for its wood's exceptional hardness and durability. Loggers exploited it intensively during the timber boom era in the Pastaza province.
- **Ice-cream bean (*Guaba bejuca*) — *Inga edulis* (FABACEAE):** A fast-growing leguminous tree with abundant biomass.

Omar Tello's restoration methodology uses it as a pioneer species to recover degraded and eroded soils. Its leaves add nitrogen to the soil. Restorers plant it in high density and manage it with periodic pruning until native forest species establish themselves.

- **Kili palm (*Kili*) — *Wettinia maynensis* (ARECACEAE):** A characteristic palm of humid Amazonian forests, present as a relict species on the land chosen for restoration. Its specimens, over fifty years old, proved fundamental for seed collection at the project's inception.
- **Laurel (*Laurel*) — *Cordia alliodora* (BORAGINACEAE):** A timber tree frequently used in the Amazon and other tropical regions. Its light-colored, easily workable wood faced widespread exploitation for construction and carpentry.
- **María panga (*María panga*) — *Piper umbellatum* (PIPERACEAE):** A medicinal and food plant of ancestral use in the Ecuadorian Amazon. Indigenous traditional medicine uses its large leaves to wrap food and as an anti-inflammatory.
- **Melastomataceae (*Melastomataceae*) — Botanical family:** A family of flowering plants, abundant in American tropical forests. In restored ecosystems, its species appear naturally thanks to bird and wind dispersal, enriching the recovering forest's floristic diversity.
- **Moral fino / Moral bobo (*Moral fino / Moral bobo*) — *Clarisia racemosa* (MORACEAE):** An Amazonian timber tree whose wood people prized for its durability. Loggers exploited it intensively in different Amazonian zones during the 20th-century timber boom.
- **Morete palm / Buriti (*Morete*) — *Mauritia flexuosa* (ARECACEAE):** A palm holding great ecological and cultural importance in the Amazon. It grows in flood-prone areas and provides a source of food, fiber, and oil. The restoration methodology recommends this species, noting restorers must plant it in groups with sufficient spacing due to its expansive leaves.
- **Naranjilla / Golden seed (*Naranjilla / Pepa de oro*) — *Solanum quitoense* (SOLANACEAE):** A tropical fruit with intense flavor and high nutritional value, popularly known as *pepa de oro* (golden seed) due to its profitability in the 1950s–

1970s. Cultivating it intensively in monocultures entailed massive primary forest clearing. According to the author, the sudden, simultaneous appearance of pests and diseases across all crops served as 'the second lesson nature taught' regarding the dangers of monoculture.

- **Pambil palm (*Pambil*) — *Socratea exorrhiza* (ARECACEAE):** A characteristic Amazonian palm, recognizable by its stilt roots. It represents one of the few specimens found on the original "Las Orquídeas" Botanical Garden land at the start of restoration.
- **Papa tuyu (*Papa tuyu*) — *Dioscorea sp.* (DIOSCOREACEAE):** A wild and cultivated tuber used for food in the Ecuadorian Amazon. It represents Indigenous knowledge regarding the forest's edible resources.
- **Papangu (*Papangu*) — *Cyclanthus bipartitus* (CYCLANTHACEAE):** A plant used for medicine and food in the Ecuadorian Amazon. Its presence in the forests reflects ancestral Indigenous knowledge concerning local flora properties.
- **Peach palm (*Chontaduro*) — *Bactris gasipaes* (ARECACEAE):** A cultivated palm holding great dietary importance for Amazonian Indigenous peoples. Its fruits are rich in carbohydrates, proteins, and fats. It forms part of the traditional crops in Indigenous *chacras.*
- **Peruvian mahogany (*Caoba / Caoba peruana*) — *Platymiscium pinnatum* (FABACEAE):** A fine wood tree with high historical commercial demand in the Ecuadorian Amazon. Its intensive exploitation during the 20th century contributed to its threatened species status.
- **Pigüe (*Pigüe*) — *Piptocoma discolor* (ASTERACEAE):** A fast-growing pioneer tree that naturally colonizes eroded soils and deforested areas. Experts consider it an indicator of environmental degradation. Although it provides some biomass during its first decades, its roots rot upon reaching maturity (between 20 and 30 years), and falling trees can damage planted native species. The restoration methodology recommends its early elimination in areas slated for restoration.
- **Pilche / Chanúl (*Pilche / Chanúl*) — *Humiriastrum procerum* (HUMIRIACEAE):** A high-density, durable timber

tree, exploited during the timber boom era in the Ecuadorian Amazon. It currently remains scarce in intervened forests.

- **Piperaceae (*Piperaceae*) — Botanical family:** A tropical plant family that includes shrubs, lianas, and herbs. In restored ecosystems, some members appear naturally and fulfill important functions in the understory, serving as food and shelter for various fauna species.
- **Plantain (*Plátano*) — *Musa spp.* (MUSACEAE):** A fundamental food crop in Indigenous and Amazonian colonizer *chacras*. It stood as one of the first products Indigenous communities traded with towns. It remains a subsistence crop in areas with better quality soils.
- **Rhinoceros beetle (*Escarabajo rinoceronte*) — *Dynastes sp.* (SCARABAEIDAE):** A large beetle, emblematic of Amazonian biodiversity. The author remembers it as a childhood toy, a symbol of the insect abundance that characterized the Amazon before massive deforestation. Its current rarity serves as an indicator of ecological deterioration.
- **Rolaquimba (*Rolaquimba*) — *Vasconcella monoica* (CARICACEAE):** A wild Amazonian plant with dietary use. It forms part of the region's Indigenous peoples' ancestral botanical knowledge.
- **Rubiaceae (*Rubiaceae*) — Botanical family:** One of the most diverse flowering plant families in the tropics. In the Ecuadorian Amazon, it includes trees, shrubs, and medicinal plants like *uña de gato* (*Uncaria tomentosa*). Its members actively participate in the ecological relationship network between flora and fauna in the restored ecosystem.
- **Tachuelo (*Tachuelo*) — *Zanthoxylum sp.* (RUTACEAE):** A spiny Amazonian tree, conserved on the original land for its potential ecological value. Although it does not produce fine wood, the author considered preserving it important, anticipating it might face the risk of disappearing in the future.
- **Tamburo (*Tamburo*) — *Vochysia sp.* (VOCHYSIACEAE):** An Amazonian tree conserved on the original restoration land based on preventive ecological criteria. Locals use its wood in rural construction.
- **Taro / Papa china (*Papa china*) — *Xanthosoma spp.* (ARACEAE):** A traditional crop tuber in Amazonian

Indigenous *chacras* and in intensive farming as a source of economic income in the region. It provides an important source of carbohydrates for the communities.

- **Ungurahua palm (*Ungurahua*) — *Oenocarpus bataua* (ARECACEAE):** A palm producing a highly nutritious oleaginous fruit, holding great ecological and dietary importance in the Amazon. Birds, mammals, and humans consume its fruits. The restoration methodology recommends it for its value as a food source for wildlife.
- **Wild mountain papaya / Mountain cabbage (*Col de monte*) — *Vasconcella microcarpa* (CARICACEAE):** A wild plant whose leaves colonizer and Indigenous families gathered for food. Its presence indicates the richness of spontaneous Amazonian vegetation, which provided food without requiring formal cultivation.
- **Yellow poui / Guayacán (*Guayacán*) — *Handroanthus chrysanthus* (BIGNONIACEAE):** A tree with spectacular flowering, recognizable by its yellow flowers that appear before its leaves. It requires at least three hundred years to produce high-quality wood. When the restoration of the "Las Orquídeas" Botanical Garden began, only one specimen existed on the land; the author considered its presence highly important for the process.

II. Ecological and Technical Terms

- **Agroforestry / Agroforestry system (*Agroforestal / Sistema agroforestal*):** A land-use modality that integrates trees and shrubs with agricultural crops or livestock in the same space. In the Amazonian context, the author proposes agroforestry systems as a sustainable alternative to intensive monoculture, due to their capacity to maintain biodiversity, protect the soil, and generate community economies without destroying the ecosystem.
- **Biodiversity (*Biodiversidad*):** The variety of life forms in a given ecosystem, including species, genetic, and ecosystem diversity. The book uses it as a central concept to highlight the wealth lost in the Ecuadorian Amazon due to colonization, extractivism, and monocultures, and sets it as the main goal of ecological restoration processes.

- **Biological corridor (*Corredor biológico*):** A strip of vegetation that connects two or more isolated habitat patches, allowing fauna to move between them. The proximity or absence of biological corridors acts as a determining factor when planning an ecological restoration project, as it directly influences the chances of wildlife returning.
- **Biological succession / Ecological succession (*Sucesión biológica / Sucesión ecológica*):** The natural and gradual process by which a community of organisms in an ecosystem progressively replaces another, from pioneer species to stable climax communities. The author's restoration methodology bases itself on a deep analysis of biological succession processes, artificially accelerating what nature would take centuries to achieve on its own.
- **Biomass (*Biomasa*):** The total organic matter living beings produce in an ecosystem, including leaves, branches, roots, and fallen trunks. In the restoration methodology, the progressive accumulation of plant biomass is key to regenerating the topsoil, incorporating nutrients into the soil, and establishing adequate conditions for fauna.
- **Climate change (*Cambio climático*):** The long-term alteration of global climate patterns, intensified by human greenhouse gas emissions and deforestation. The book documents its impact in the Amazon: extreme droughts (such as the one in 2010), altered species flowering and fruiting cycles, dropping river levels, and the destabilization of pollinator and wildlife communities.
- **Destroyed environment (*Entorno destruido*):** A category proposed by the author to designate spaces where human activity completely eliminated original vegetation and replaced it with monocultures or other uses that prevent the forest's spontaneous regeneration. The typical example is cattle pasture, where human intervention remains permanent. These represent priority spaces for assisted restoration.
- **Ecological restoration (*Restauración ecológica*):** The assisted process of recovering a degraded, damaged, or destroyed ecosystem. Within the book's framework, ecological restoration does not limit itself to reforestation using commercial species. Instead, it seeks the ecosystem's comprehensive reconstruction: diversified native flora,

including threatened species; free-ranging wildlife; and links to Indigenous cultural knowledge. Omar Tello's central research hypothesis states: 'If you plant the extinct flora in a certain place, then the lost fauna will return.'

- **Ecosystem (*Ecosistema*):** A system formed by a specific area's living organisms and their environment's physical elements (soil, water, climate), in constant interaction. The concept remains central to the book: restoration's goal does not simply mean planting isolated trees, but rather rebuilding a functional ecosystem that integrates its forest, biological, and cultural components.
- **Epiphyte (*Epífita*):** A plant that grows on another—generally on tree branches or trunks—without parasitizing it, using it solely as support. Orchids, bromeliads, and some heliconias are characteristic Amazonian epiphytes. In the book's context, the text documents the massive extraction of epiphytes by international collectors in the 1970s as one of the first acts of plundering Amazonian biodiversity.
- **Extractivism (*Extractivismo*):** A natural resource exploitation model based on the intensive extraction and export of raw materials (wood, oil, minerals, fauna, wild flora) without significant processing or planning for resource renewal. The author identifies extractivism as the structural cause of the Ecuadorian Amazon's destruction, encompassing everything from timber logging and commercial hunting to oil exploitation and orchid extraction.
- **Girdling / Debarking (*Anillado / Descortezado*):** A forestry technique that involves removing a ring of bark around a tree trunk to cause its progressive death. In secondary forest restoration, restorers apply it to unwanted trees to reduce their impact when falling: by dying slowly, they decompose gradually and add nutrients to the soil with less risk of damaging surrounding seedlings.
- **Humanized environment (*Entorno humanizado*):** A category proposed by the author for forests that, despite appearing natural, have suffered selective intervention through the extraction of fine woods or other species. It includes secondary forests and reforestation projects aimed at timber harvesting. In these environments, species diversity remains incomplete and the ecosystem's structure is altered.

- **Living germplasm bank (*Banco de germoplasma vivo*):** A collection of living plants maintained in an open field aimed at conserving the genetic diversity of threatened or endangered species. Unlike conventional seed banks, the "Las Orquídeas" Botanical Garden's living germplasm bank allows for continuous monitoring, periodic seed collection, and propagation for distribution to communities and restoration projects.
- **Macrofauna (*Macrofauna*):** In Tello's terminology, the larger animals whose return can be readily observed, including mammals, birds, reptiles, and larger amphibians. The usage reflects the author's field classification rather than a formal body-size category in some ecological disciplines.
- **Micro-watershed (*Microcuenca*):** A small hydrographic unit that captures and drains rainwater into a main channel. The text points to the destruction of micro-watersheds due to deforestation as one of the gravest consequences of IERAC policies, as it affects the water supply of entire communities.
- **Microfauna (*Microfauna*):** In Tello's terminology, the smaller animal life of the ecosystem, particularly insects, spiders, and other small organisms whose progressive return he documented. The term is used more broadly here than in some technical ecological classifications, where "mirofauna" may refer specifically to microscopic or very small animals.
- **Monoculture (*Monocultivo*):** The agricultural practice of growing a single plant species over large tracts of land. The author documents how monocultures of naranjilla, cattle pasture, African palm, and other products promoted in the Amazon triggered the appearance of pests, soil depletion, and chemical pollution, in what he describes as the 'lessons of nature' that society failed to read.
- **Natural environment (*Entorno natural*):** A category proposed by the author for ecosystems that human activity has never significantly intervened in or polluted. In these spaces, ecological processes operate without interference, and the recovery dynamic following natural disturbances remains positive. In these environments, only collecting seeds for propagation programs is justifiable; extracting plants is not.
- **Pasture / Pastureland (*Pastizal*):** A tract of land covered with cultivated grasses to feed livestock. Converting primary

forests into pastures, mainly using *gramalote* (*Axonopus scoparius*), constituted the principal form of forest destruction in the Ecuadorian Amazon during the colonization decades. The original "Las Orquídeas" Botanical Garden land consisted of a 6.75-hectare pasture with eroded soils when restoration began in 1980.

- **Phenology (*Fenología*):** The study of organisms' periodic biological cycles and their relationship with climate and seasons. In the climate change-affected Amazon, the phenology of many species has altered: trees that previously produced seeds annually now do so every two, three, or four years, disrupting the feeding schedules of the fauna that depend on those fruits.
- **Pollinator (*Polinizador*):** An animal that transports pollen from one flower to another, enabling plant sexual reproduction. In the Amazon, pollinators include hummingbirds, butterflies, bees, beetles, and bats. Their return to the restored ecosystem acts as a fundamental indicator of ecological recovery, as it allows trees and plants to begin producing seeds naturally.
- **Primary forest (*Bosque primario*):** A forest that human activity has not significantly disturbed, characterized by high native species diversity, large trees, and a complex vertical structure. In the text, it serves as the reference point for historical abundance and the standard against which one compares the restoring ecosystem.
- **Secondary forest (*Bosque secundario*):** A forest that grows after a significant disturbance to the original forest, whether from logging, fire, or other causes. It presents lower species diversity than primary forest and features dominant fast-growing pioneer species. The restoration methodology approaches secondary forests using a strategy of complementing and selectively replacing species.
- **Seedling (*Plántula*):** A young plant in its early development stages, following seed germination. Obtaining, transporting, and establishing seedlings of threatened species, rescued from logged forests undergoing destruction, formed one of the core activities in the restoration methodology developed by Omar Tello.

- **Selective thinning (*Raleo selectivo*):** The practice of controlling and planning the removal of some trees in a forest to reduce competition for light and space, thereby favoring the growth of other species with higher ecological value. In secondary forest restoration, thinning provides an essential tool to open clearings that allow restorers to incorporate new species.
- **Topsoil / Soil organic layer (*Capa vegetal / Capa orgánica del suelo*):** The superficial soil horizon formed by accumulating and decomposing organic matter. In the colonized Amazon's degraded, clayey soils, this layer had almost completely disappeared. Reconstructing it stood as the restoration project's first technical challenge, which the author tackled by massively recycling sawdust from timber factories and sugarcane bagasse, mixing them with animal organic matter.
- **Trophic chain / Food chain (*Cadena trófica / Cadena alimentaria*):** A sequence of feeding relationships through which energy and nutrients pass among organisms. Tello frequently uses the term more broadly to describe the interconnected feeding relationships of the ecosystem, what ecology may also describe as a food web. The author uses this concept to show that Indigenous communities formed an integral part of the Amazonian food chain, and that breaking any of its links—whether through species extinction, deforestation, or cultural displacement—triggers an imbalance that affects the entire system.
- **Wetland (*Humedal*):** An area where water permanently or temporarily saturates the soil, such as swamps, bogs, and riverbanks. In the restoration methodology, restorers treat wetlands as key ecosystem health indicators: the quality and quantity of water they accumulate reflects the surrounding forest's recovery degree. Reforesting them with aquatic plants forms part of the restorative process.

III. Cultural and Historical Terms

- **Achuar (*Achuar*):** An Indigenous nationality of the Ecuadorian Amazon, settled primarily in the Pastaza province and regions bordering Peru. It forms part of the seven nationalities that, according to the author, currently inhabit the Pastaza province. It maintains a close relationship with the region's river and forest ecosystems.

- **Agrarian Reform Law (*Ley de Reforma Agraria*):** The Ecuadorian legal framework that established conditions for land redistribution in the country. The first Agrarian Reform Law dates back to 1964 (Decree 1480 of the Military Government Junta) and saw reforms in 1973 (Decree 1172). In practice, in the Amazon, it functioned as an instrument that legalized massive deforestation by demanding the land's 'productivization' as a requirement for issuing deeds.
- **Amazonian Quichua / Kichwa (*Quichua amazónico / Kichwa amazónico*):** The Indigenous nationality of the Ecuadorian Amazon with the highest number of speakers in the region. Various groups adopted its language, a variant of Andean Quechua, during historical regional integration processes. It is one of the seven Pastaza province nationalities mentioned in the book.
- **Ancestral knowledge (*Saberes ancestrales*):** The knowledge, practices, techniques, and worldviews Indigenous peoples accumulated over generations, transmitted orally from parents to children. In the Amazonian realm, they include detailed knowledge of thousands of plants' medicinal properties, fauna behavioral patterns, hydrological cycles, and diversified farming techniques. The author considers this knowledge to constitute 'the greatest strength of all Indigenous cultures' and views its dispossession as one of cultural extractivism's gravest acts.
- **Andoa (*Andoa*):** An Indigenous nationality of the Ecuadorian Amazon with a small number of speakers. It inhabits the Pastaza province and experts consider it one of the cultures at highest risk of disappearing in the region. The author includes it among the seven nationalities of Pastaza whose ancestral science faces the threat of going extinct alongside its last wise elders.
- **Barbasco (*Barbasco*):** A name applied to several Amazonian plants traditionally used in fishing. Some contain rotenone or related compounds that interfere with cellular respiration in fish, temporarily stunning or killing them and allowing them to be collected.
- **Biocentrism (*Biocentrismo*):** An ethical position that recognizes intrinsic value in all living beings, independent of their usefulness to humans. It remains implicit in the philosophy supporting Omar Tello's entire work: respecting

the life of every insect, bird, or plant justifies itself not by economic utility, but by its right to exist. Appendix A develops this as a conceptual framework in relation to the Rights of Nature jurisprudence enshrined in Ecuador's 2008 Constitution.

- **Chacra (*Chacra*):** A small domestic or family farming plot used by Amazonian Indigenous peoples. It combines various edible species such as plantain, cassava, taro, peach palm, and barbasco in a productive system that respects the surrounding ecosystem's equilibrium. The *chacra* represents a diversified agriculture model opposed to intensive monoculture.
- **Colonization / Colonization process (*Colonización / Proceso de colonización*):** An Ecuadorian state policy implemented since the 1960s, through which the government incentivized peasant families to settle in the Amazon by granting land. To legalize ownership, families had to clear at least 80% of the forest. The author identifies this process as the direct cause of massive forest destruction, the cultural deterioration of Indigenous nationalities, and the beginning of the Amazonian ecological imbalance.
- **Ecuadorian Institute of Agrarian Reform and Colonization (*IERAC - Instituto Ecuatoriano de Reforma Agraria y Colonización*):** An Ecuadorian state institution created by Decree 1480 on July 11, 1964, tasked with implementing the Agrarian Reform and Colonization Law. It distributed 50-hectare plots of virgin jungle to families who committed to clearing at least 80% of the forest to prove 'productive use' of the land. The author brands it 'the greatest culprit of the *minifundio* and deforestation in Ecuador,' for having institutionalized the destruction of primary forests as a legal condition for land adjudication. The government extinguished and replaced it with other institutions in subsequent years.
- **Environmental justice (*Justicia ambiental*):** A principle demanding an equitable distribution of environmental benefits and burdens among different social groups, without discrimination based on ethnic origin, social class, or geographic location. Appendix A applies it to analyze how Amazonian Indigenous peoples have historically borne the

environmental costs of extractivism without receiving its benefits.

- **Feminist political ecology (*Ecología política feminista*):** An academic school of thought that analyzes the relationships between gender, power, nature, and territory. The book's Appendix A utilizes this theoretical framework to enrich the reading of Omar Tello's testimony, pointing out how patriarchal and extractivist power structures disparately affect Amazonian communities and their environments.
- **Indigenous epistemology (*Epistemología indígena*):** A set of ways of knowing, interpreting, and relating to the world specific to Indigenous peoples, built through generations of observing and coexisting with the natural environment. The author argues that the elders of Amazonian Indigenous communities possessed an ancient science regarding plant properties, fauna cycles, and ecosystem balances, and that colonizers and evangelizers deliberately ignored and stripped away this science.
- **Minifundio / Smallholding (*Minifundio*):** The excessive division of land into plots too small for profitable and sustainable agricultural production. The author points out that IERAC policies generated the Amazonian *minifundio* by distributing land without technical or sustainability criteria, contributing to territorial fragmentation and the socioeconomic deterioration of colonizer families.
- **Plurinational and Intercultural State (*Estado Plurinacional e Intercultural*):** A constitutional concept Ecuador adopted in 2008 that recognizes the diversity of nations and cultures coexisting within the national territory. The author acknowledges the progress this represents on paper, but notes that in practice, Amazonian communities remain marginalized, lacking equitable access to health, education, and territorial justice.
- **Rights of Nature (*Derechos de la Naturaleza*):** The legal recognition of nature as a subject of rights, enshrined in Ecuador's 2008 Constitution. Despite this formal protection, the author documents how state institutions continued to threaten his restoration project even after that date, ignoring the constitutionally recognized rights of Amazonian flora and fauna.

- **Sacha runa (*Sacha runa*):** An expression in the Quichua language that literally means 'jungle man' or 'forest being'. In the Amazonian Indigenous worldview, the *sacha runa* is a spiritual or supernatural being that inhabits the deep jungle. Don Camilo Santi, an Indigenous elder the author interviewed, explained that a black dog could alert a traveler to the *sacha runa*'s presence during night treks.
- **Shiwiar (*Shiwiar*):** An Indigenous nationality of the Ecuadorian Amazon, settled in the Pastaza province. It maintains strong ties with the region's forest and river ecosystems. It forms one of the seven nationalities the author mentions as current inhabitants of the province.
- **Shuar (*Shuar*):** An Indigenous nationality of the Ecuadorian Amazon, one of the most numerous in the region. Historically known for its warrior organization and resistance to colonization. The book mentions it as part of the mosaic of Amazonian cultures whose way of life and knowledge environmental destruction and colonization processes have affected.
- **Slow violence (*Violencia lenta*):** A theoretical concept developed by academic Rob Nixon to describe forms of violence that occur gradually and imperceptibly over time, lacking the drama of a singular violent event. Appendix A applies it to the analysis of Amazonian destruction: progressive deforestation, river pollution, the gradual extinction of species, and the cultural dispossession of Indigenous peoples constitute forms of slow violence that fail to generate the media urgency they deserve.
- **Testimonio / Literary genre (*Testimonio*):** A Latin American narrative genre in which an individual voice relates a collective experience of marginalization, resistance, or struggle. Appendix A, written by editor Jon Dell Jaramillo, situates Omar Tello's work within this tradition: a testimony that speaks not only for its author, but for the jungle, Indigenous communities, and all beings lacking their own voice in the political debate.
- **The Ecuadorian Oriente / The Ecuadorian East (*El Oriente Ecuatoriano*):** A historical and popular denomination used to refer to the Ecuadorian Amazon before authorities administratively recognized the region as an integral part of the country. The expression reflects the cultural and political

distance that separated the Amazon region from the central Andean power, and the widespread ignorance regarding its biological and cultural wealth.

- **Waorani (*Waorani*):** An Indigenous nationality of the Ecuadorian Amazon, known for its historical resistance to contact with the outside world and for having inhabited the deep interior of the jungle for centuries. It stands as one of the seven nationalities mentioned as present in the Pastaza province. It has been one of the communities most severely affected by oil exploitation in its territory.
- **Worldview / Cosmovision (*Cosmovisión*):** A system of beliefs, values, and perceptions through which a people interprets the world and its place within it. The worldviews of Amazonian Indigenous nationalities recognize a continuous relationship between human beings, nature, and the spiritual world, without establishing a hierarchy of domination over the environment. The author points out that destroying this worldview, through evangelization and colonization, constituted one of the civilizatory process's most irreparable losses.
- **Zápara (*Zápara*):** An Indigenous nationality of the Ecuadorian Amazon, which UNESCO declared an Intangible Cultural Heritage of Humanity in 2001 for its language and oral traditions. It represents one of the cultures at highest risk of linguistic and cultural extinction in the region. The author includes it among the seven nationalities of Pastaza whose survival remains threatened.

INDEX